KU-517-169

Chemical Periodicity

MODERN CHEMISTRY BACKGROUND READERS

Edited by **J. G. Stark**
Head of Chemistry, Glasgow Academy

Other titles in the series:

* The Shapes of Organic Molecules *N. G. Clark*
Electrochemistry *P. D. Groves*
Inorganic Complexes *D. Nicholls*
* Structure in Inorganic Chemistry *E. Sherwin*

* *In preparation*

Chemical Periodicity

D. G. Cooper BSc FRIC
Head of Science, Birkenhead College of Technology

John Murray Albemarle Street London

Text set in 10/11 pt. Monotype Times New Roman, printed by photolithography, and bound in Great Britain at The Pitman Press, Bath

0 7195 2934 4

Contents

1 Periodicity and Atomic Structure

Science consists of an *orderly* body of knowledge, and a part of the orderliness is the convenient division of science into branches such as physics, biology, and chemistry. None of these branches made much progress until systematic relationships had been recognized; these relationships may or may not have been mathematical, but those which did involve mathematics seem to have been recognized quite early, as in the cases of astronomy and mechanics. Once a reasonable body of factual knowledge existed, it often happened that two or more people more or less simultaneously recognized a system or pattern, connecting the facts and giving some explanation of them.

There have been several examples of simultaneous discovery in chemistry, and one occurred a little over a hundred years ago when the Englishman, Newlands, proposed his *Law of Octaves* (1864) at about the same time that Mendeleev, a Russian chemist, published his *Periodic Law* (1872). At the time, Newlands's theory suffered because he chose music as an analogy. However, we can now appreciate that the analogy was not at all a bad one. In it, Newlands compared the recurrence of a musical note at the same position in each octave, with the recurrence of chemical properties in every eighth element: but his work received no acknowledgement until after his death. Mendeleev dealt with more elements and his law was put to almost immediate use; as a result he completely overshadowed Newlands. They both arranged the known elements in order of increasing atomic weight (now called 'relative atomic mass') and compared their properties, to show that there was a pattern of recurring similarities. This system could not have been discovered earlier, because the atomic weights, even of the common elements, were only just being determined (and even then were not always accurate).

Mendeleev realized that more elements remained to be discovered and boldly left gaps for these in his periodic table. He listed the properties of several elements then unknown, which he 'determined' in considerable detail by taking an average of the known elements on either side of the gaps. In some cases he suggested that the accepted values for atomic weights were inaccurate. He thus provided the stimulus for further work, as a result of which the missing elements turned up quite soon, and some atomic weights were corrected. It was this predictive function of the Periodic Law which established its early reputation.

Some 25 to 30 years after Mendeleev published his law, when the predictive function of the law was becoming less important, a second function began to take over. By this time the electron had been discovered, and the inert or noble gases had been isolated (beginning in 1895). The desire to discover the reason for the recurring pattern of the properties of the elements led to a consideration of the structure of atoms—a function which has proved of greater importance than the first. The periodic table played a large part in the development of atomic theory and the electronic theory of valency; and an understanding of the shell structure of electrons round atoms immediately explained the form of the periodic table. Thus we have an important fundamental development arising directly from systematic classification.

The usual form of the periodic table shows periods (usually arranged horizontally) of elements whose properties change gradually, and groups (arranged vertically); all elements falling in a given group have considerable similarity to each other. It is now understood that the elements in a group have the same outer electron configuration, while the elements in a period show a systematic variation of the electrons in a shell, or in adjoining shells. In general, successive periods represent the addition of successive shells of electrons.

What might be called the architecture of periodic tables has varied through the years, with particular emphasis being placed either on the practical usefulness of the tables or on aspects of atomic structure. Many people have been concerned with these variations, not all of which have much significance or lasting value. There were attempts to emphasize the continuous nature of the periods, some of which led to three-dimensional structures. Picture a multi-stage rocket with a spiral line drawn on its surface: the nose cone is hydrogen–helium; the top stage has the two short periods; then, after the bulge to the next stage, we have the first and second long periods; and so on. An attempt to give a similar impression using only two dimensions gave a 'race course' layout. A comparison of textbooks of different publication dates will probably reveal some differences in the arrangement of periodic tables; the one reproduced here (figure 1) is in accordance with the most recent convention. Since it is not found in identical form in all books at this level, it deserves a little explanation.

The terms 's-block', 'p-block', 'd-block', and 'f-block' in the table refer to the electron sub-shells which are being filled in the atoms shown in the block. It has already been mentioned that successive periods are associated with successive shells of electrons, and the letters s,p,d, and f are used to classify the types of orbitals that electrons occupy within a shell (see table 1). (The letters originate from the terms *s*harp, *p*rincipal, *d*iffuse, and *f*undamental, names applied to spectroscopic lines.) The first shell cannot be divided into sub-shells because it only contains one orbital, the 1s orbital, which can accommodate a maximum of two electrons, which must be of opposite spin. Each orbital, in whatever

	s-block		d-block										p-block						Atomic numbers
Group	I	II											III	IV	V	VI	VII	VIII	
	Li	Be											B	C	N	O	F	Ne	3–10
	Na	Mg											Al	Si	P	S	Cl	Ar	11–18
	K	Ca	Sc	Ti	V	Cr	Mn	Fe	Co	Ni	Cu	Zn	Ga	Ge	As	Se	Br	Kr	19–36
	Rb	Sr	Y	Zr	Nb	Mo	Tc	Ru	Rh	Pd	Ag	Cd	In	Sn	Sb	Te	I	Xe	37–54
	Cs	Ba	La[1]	Hf	Ta	W	Re	Os	Ir	Pt	Au	Hg	Tl	Pb	Bi	Po	At	Rn	55–86
	Fr	Ra	Ac[2]																87–89
				f-block															
				[1]Ce	Pr	Nd	Pm	Sm	Eu	Gd	Tb	Dy	Ho	Er	Tm	Yb	Lu		58–71
				[2]Th	Pa	U	Np	Pu	Am	Cm	Bk	Cf	Es	Fm	Md	No	Lr		90–103

Figure 1 A modern form of the periodic table.

sub-shell, can accommodate two electrons, as long as they are of opposite spin. In the second shell, there are s and p sub-shells, the first of which has one orbital (capable of holding two electrons), and the second three orbitals (six electrons). The third shell adds a further five orbitals with room for ten d electrons, so that there are 3s, 3p, and 3d orbitals, while the fourth adds seven more orbitals which can hold fourteen f electrons.

Shell	Orbitals				Total	Total
	s	p	d	f	orbitals	electrons
1	1	0	0	0	1	2
2	1	3	0	0	4	8
3	1	3	5	0	9	18
4	1	3	5	7	16	32

Table 1

At one time, according to the Bohr theory, electrons were considered to travel within a shell in definite orbits rather as if they were on invisible railway lines. The word 'orbit' was replaced by the term 'orbital' at the same time that this simplified theory was discarded. Electrons are now thought of in terms of probability: they 'occupy' a diffuse zone whose boundaries are defined by the solution of a mathematical probability equation. The physical shapes of the orbitals are not of great importance at this stage; but it may be of interest to note that s orbitals are spherical, while p orbitals are dumb-bell-shaped and have directions in space which are mutually perpendicular (*x*-, *y*-, and *z*-axes).

Electron configurations of the elements of the first short period

The first period, corresponding to the first shell of electrons, contains only the two elements hydrogen and helium, and the name 'first short

Atomic number	Element	1s	2s	2p		
3	lithium	2	↑			
4	beryllium	2	↑↓			
5	boron	2	↑↓	↑		
6	carbon	2	↑↓	↑	↑	
7	nitrogen	2	↑↓	↑	↑	↑
8	oxygen	2	↑↓	↑↓	↑	↑
9	fluorine	2	↑↓	↑↓	↑↓	↑
10	neon	2	↑↓	↑↓	↑↓	↑↓

Figure 2 Electron configurations of the elements Li to Ne.

period' is usually given to the elements 3 to 10. The orbitals for these may be illustrated as in figure 2 with each box representing an orbital and each arrow representing an electron. The direction of the arrows indicates the opposite spin of the two electrons which pair off in each orbital (the 1s orbital is completely filled throughout and is, therefore, omitted).

It will be noticed that carbon and nitrogen atoms have two and three electrons respectively in p orbitals but not paired. Pairing only starts after each of the available orbitals has one electron in it, that is the number of electrons with unpaired spin is a maximum (*Hund's maximum multiplicity rule*).

Electron configurations of the elements in the later periods

The numbers of electrons which can be accommodated in successive shells account for the numbers of elements in the periods:

2s + 6p = 8, as in the short periods
2s + 6p + 10d = 18, as in the first and second long periods
2s + 6p + 10d + 14f = 32, as in the third long period

These totals are not found in successive periods—the totals are 2, 8, 8, 18, 18, 32—because the order in which orbitals are filled by electrons depends on the energy levels: the orbitals first filled are of lowest energy, and the filling takes place in order of increasing energy level as shown in figure 3. The process starts as one would perhaps expect with 1s, 2s, and 2p orbitals being filled as shown in the first short period (above); then the 3s and 3p orbitals fill in the corresponding elements of the second short period (sodium to argon). However, the next to be filled after the 3p orbitals are not the 3d orbitals but the 4s which happen to be of lower energy than the 3d.

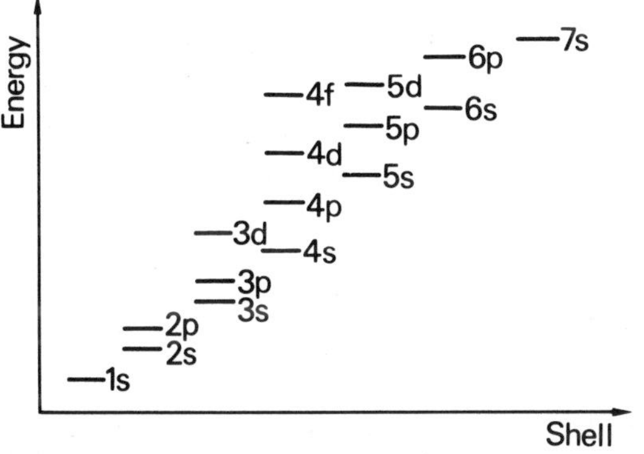

Figure 3 Energy levels of orbitals in an atom (not to scale).

This brings us to elements 19 and 20, potassium and calcium, witl one and two 4s electrons respectively; then element 21, scandium, add one 3d electron, after which the 3d orbitals are progressively fille before the 4p orbitals. Elements 21 to 29, scandium to copper, have (nearly) constant outer electron configuration, and are adding electron in an inner shell—the third shell—when there are already two electron in the fourth shell (see figure 4). Such elements have long been known a *transition elements*, and are placed in the d-block of elements. Transi tion elements usually possess certain special properties which are no found to a great extent outside their ranks. These include variabl oxidation states, coloured ions, catalytic activity, paramagnetism, and strong tendency to form complex ions (see Chapter 7). These propertie are not exclusive to transition elements, but are not often shown all a once by other elements. Neither do all the transition elements show al these properties fully developed: in scandium, for instance, they ar much less obvious than in vanadium or chromium. The point is tha transition elements are classified as such because of their electron con figurations, having incomplete electron shells other than their oute shells. They are then observed to have, in most cases, the genera properties quoted, and these have been explained in terms of atomi structure.

		third shell							fourth she
Atomic number	Element	3s	3p	3d					4s
19	potassium	2	6						↑
20	calcium	2	6						↑↓
21	scandium	2	6	↑					↑↓
22	titanium	2	6	↑	↑				↑↓
23	vanadium	2	6	↑	↑	↑			↑↓
24	chromium	2	6	↑	↑	↑	↑	↑	↑
25	manganese	2	6	↑	↑	↑	↑	↑	↑↓
26	iron	2	6	↑↓	↑	↑	↑	↑	↑↓
27	cobalt	2	6	↑↓	↑↓	↑	↑	↑	↑↓
28	nickel	2	6	↑↓	↑↓	↑↓	↑	↑	↑↓
29	copper	2	6	↑↓	↑↓	↑↓	↑↓	↑↓	↑
30	zinc	2	6	↑↓	↑↓	↑↓	↑↓	↑↓	↑↓

Figure 4 Electron configurations of the elements K to Zn. In each element th first and second shells are full, with 1s, 2s, and 2p electrons.

When we write the electron configurations of the first long perio elements, using the system given above, we find there is a large numbe of orbitals, even ignoring completed shells. For example, krypton, a the end of the first long period, has electrons occupying the following orbitals: one 3s, three 3p, five 3d, one 4s, and three 4p (in addition to the first and second shells).

As the 3s and 3p orbitals are completely filled throughout this period, we can omit these and show only the ones which may change (figure 4). Figures 2 and 4 represent the 'ground states' of the atoms, which may change during compound formation. It will be noted that chromium takes a 4s electron into the 3d shell, which is then half full, and copper also takes a 4s electron to complete the filling of the 3d shell. There seems to be an energy 'bonus' when a shell is either half full or completely full. (It should be noted that, in the case of chromium, Hund's maximum multiplicity rule is satisfied.)

On examining the periodic table given, it will be seen that the d-block of elements finishes with the zinc group; however, the elements of the zinc group are not transitional in structure or character. Zinc itself has a completed 3d sub-shell, and also has the two s electrons in the fourth shell. Because of the complete 3d sub-shell, zinc does not show variable oxidation states, does not have coloured ions, and does not show particular catalytic activity. It does, however, show a marked tendency to form complex ions.

The s-block contains only two groups of elements, all metals: group I (the alkali metals), which all have a single s electron in their outer shells; and group II (the alkaline earth metals), which all have two s electrons in their outer shells. There are, in these two cases, very strong group relationships, and the gradation of properties is adequately explained by a consideration of the relative sizes of the atoms or ions, and of the charge on the resulting ions, for example, K^+, Ca^{2+}.

The p-block shows much more variation of properties than the s-block, and is the only block containing both metals and non-metals; all the elements contain the pair of s electrons, and the group III to VII elements have one to five p electrons respectively. The noble gases have the complete p sub-shell of six electrons. The s- and p-blocks of this modern periodic table emphasize the vertical groups, and the elements in these blocks are normally considered in their groups.

Within the d-block there is less scope for vertical group comparisons, although in some cases this treatment is still a fruitful exercise; the shape of the block invites a horizontal comparison. Various physical properties can, for instance, be plotted on a graph, when it will be found that most of these properties show comparatively slow changes from one end of the block to the other (see Chapter 2).

The f-block elements do not come within the scope of A-level syllabuses, but are analogous to the d-block elements. They are known as *inner transition* elements, because the f orbitals do not fill until there are some electrons in the two outer shells, that is the first row occurs when the 4f orbitals are filling, and there are 5s, 5p, and 6s electrons present. The elements in this row are known as lanthanides and the final row elements are known as actinides; the latter include transuranic elements made synthetically.

Although the electron configuration of the atoms is important, it must be remembered that it can change, for example in compound formation, and a great deal of chemistry is systematically explained by the relative ease with which changes can be brought about.

Exercises

1 Prepare a chart similar to figure 2 for elements 11 to 18.

2 Extend figure 4 to include elements 31 to 36; compare elements 37 to 54.

2 Periodicity of Properties: Physical

When we examine the properties of the elements in their periods, we find a recurring pattern whether we are dealing with chemical or physical properties. This is not surprising, because the physical nature of the atom (including its electron configuration) determines the way the atom will behave chemically.

Some physical properties are plotted in figures 5 to 12.

Atomic radius

One of the first properties used to show periodicity was atomic volume, derived from density, and plotted graphically by Lothar Meyer. Important conclusions were drawn from atomic volume relationships, which give a valid indication of the relative sizes of atoms. However, we now have available measurements of the various atomic radii derived from X-ray analysis, which were not available in Lothar Meyer's time. It is now usual to compare atoms in terms of size or radius rather than volume.

Figure 5 gives the covalent radii for the elements of the first and second short periods (ignoring H and He) and clearly illustrates the recurring pattern. The largest atom in each period is in fact the alkali metal, the atom having only one electron in its outer shell. This electron is comparatively far-ranging; but as the positive charge on the nucleus increases, each extra electron added is attracted more powerfully and held to a smaller radius. This process continues up to the halogen.

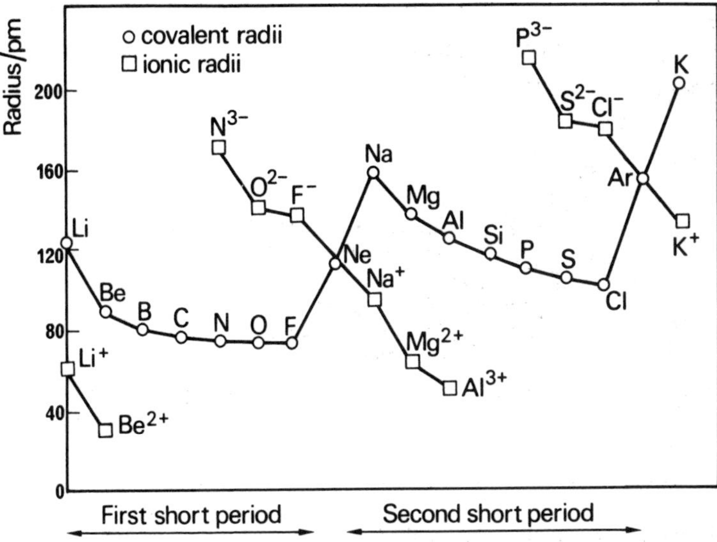

Figure 5 Covalent and ionic radii of the elements in the first and second short periods (1 pm = 10^{-12} m). (For noble gases, van der Waals radii replace covalent radii.)

The radii shown on the graph for Ne and Ar are, of necessity, not the covalent radii but the van der Waals radii, that is the radii of the uncombined, undeformed atoms. They should not strictly be compared with the covalent radii of the neighbouring elements, since the van der Waals radius is greater than the covalent radius where both can be measured.

The same graph shows the ionic radii of elements where these are significant. It will be noted that anionic radii are always considerably greater than the atomic radii of the corresponding atoms, while cationic radii are considerably less than the corresponding atomic radii. This is to be expected since the added electrons in an anion reduce the effect of the positive charge on the nucleus, and increase electron repulsion so that more space is required. On the other hand, there are fewer electrons in a cation to share the same positive charge found on the nucleus of the corresponding atom, and they are therefore 'pulled in'.

Figure 6 shows the covalent and ionic radii of the first long period elements. It will be seen that essentially there is much the same 'saucer'-shaped graph, but superimposed on the main shape is a secondary 'saucer' due to the d-block elements. The elements from vanadium to copper hardly vary at all in covalent radius: one of the features of transition metals is that they have compact atoms which can sometimes be interchanged (for example, alloys over a wide range of compositions).

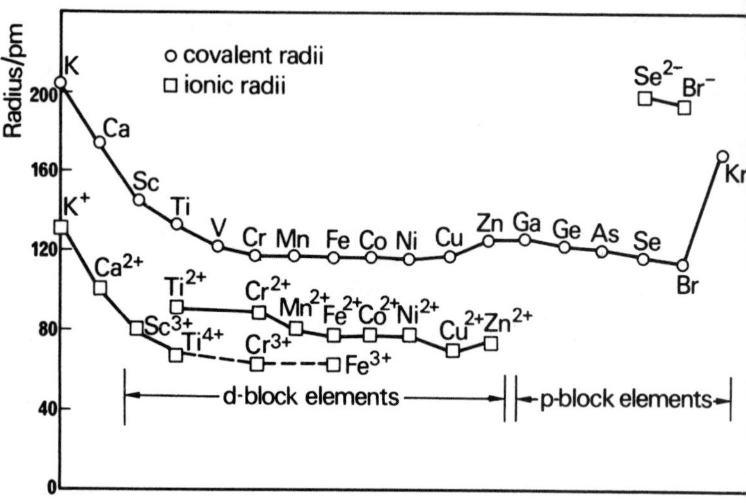

Figure 6 Covalent and ionic radii of the elements in the first long period. (Fo noble gases, van der Waals radii replace covalent radii.)

Density

Figures 7 and 8 show the densities of the elements in the short period and in the first long period. Here, the cyclic pattern shows a maximum

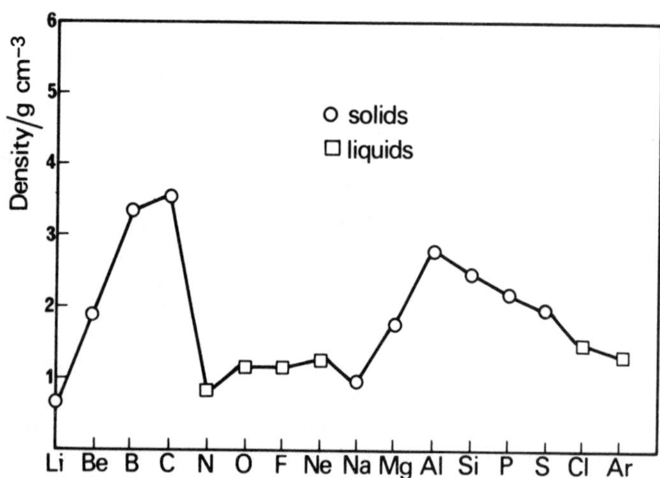

Figure 7 Densities of the elements in the first and second short periods.

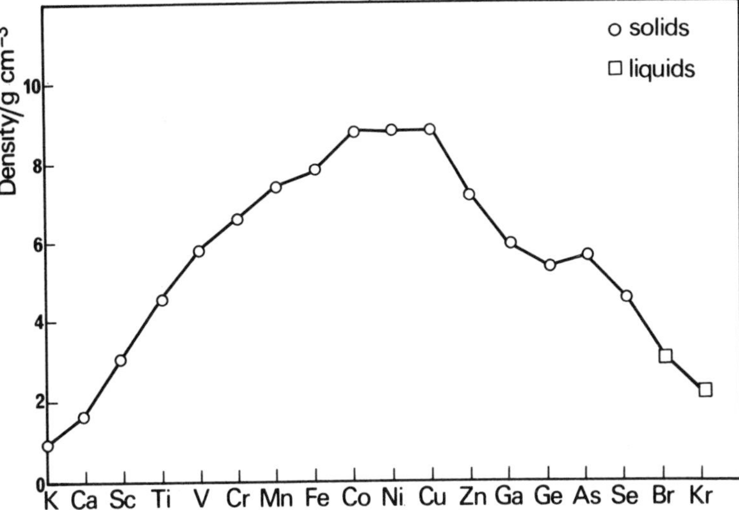

Figure 8 Densities of the elements in the first long period.

towards the centre of the period; it should, however, be remembered that the density of the element must be related to the atomic radius of the atom and to the type of crystal packing.

Boiling point and enthalpy of vaporization

Figures 9 and 10 give a similar comparison of the boiling points of the elements in the short and long periods. In this case the first short period has been omitted, because the 'permanent gases' nitrogen and oxygen have unusually low boiling points. In the long period, it will be noted that the p-block elements then repeat quite closely the graph of the short period. The boiling point of an element depends on atomic (or molecular) mass and also on the attraction that one atom has for another; since the atomic masses are only increasing slowly (in successive elements), the maxima in the graphs show that the atoms at these points must have high attraction for each other.

This is confirmed by the enthalpies of vaporization shown for comparison on the same figures. The enthalpy of vaporization of an element is the amount of energy required to separate one mole of atoms, that is to overcome their mutual attraction, in going from the liquid state to the vapour state; the atoms are still comparatively close together in the liquid state, and there is mutual attraction. It will be noted how closely the graph of enthalpies of vaporization follows that of boiling points. The minima in the long period (figure 10) correspond to manganese, in which the 3d shell is half full, zinc, in which the 3d shell is full, and

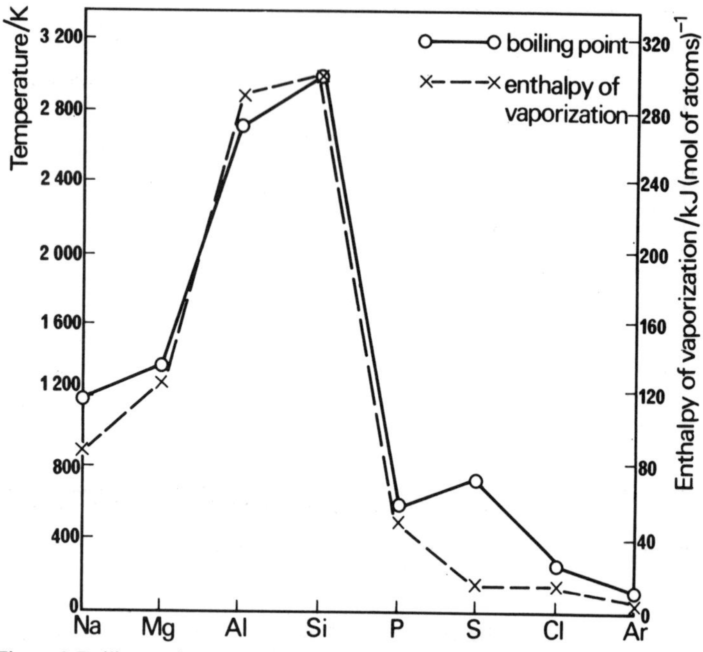

Figure 9 Boiling points and enthalpies of vaporization for the elements in the second short period.

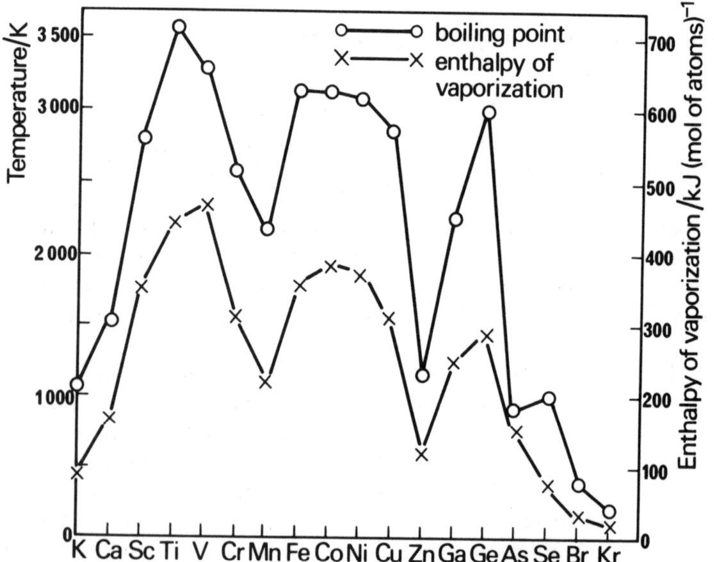

Figure 10 Boiling points and enthalpies of vaporization for the elements in the first long period.

arsenic, in which the 4p shell is half full. A similar relationship is found between melting points and enthalpies of fusion, which follow the same general shape as the boiling point graph.

Ionization energy

Figures 11 and 12 show the first ionization energies for the elements of the short and first long periods. In the case of both the short periods (figure 11), the general increase in energy required with successive elements is 'stepped' so that there are local minima in groups III and VI. Now, boron and aluminium have a single p electron, slightly easier to remove than the second s electron of the previous elements. Oxygen and sulphur have one paired p electron compared with nitrogen and phosphorus, which have three unpaired p electrons, and the one paired electron is easier to remove than one of the unpaired set. This is another way of saying that there is extra stability associated with completely filled, or half filled, sub-shells. This graph confirms spectrographic data on the distribution of electrons in their sub-shells.

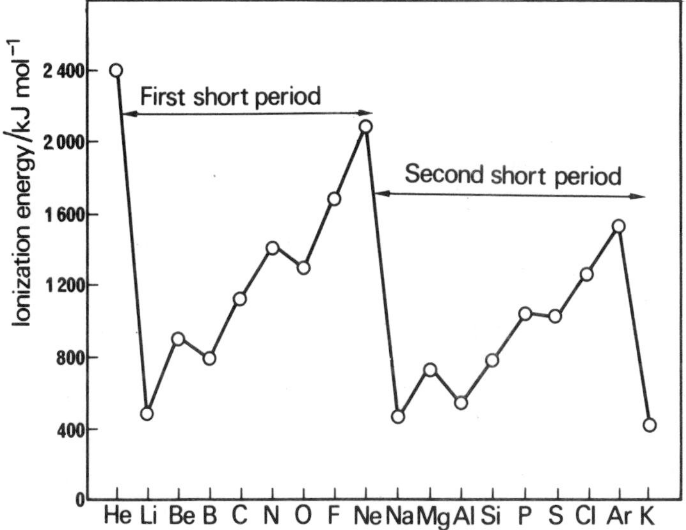

Figure 11 First ionization energies of the elements in the first and second short periods.

In figure 12, it will be noted that in the long period there is a gradual increase of ionization energy, and that the p-block elements then show the same effect as in the short periods.

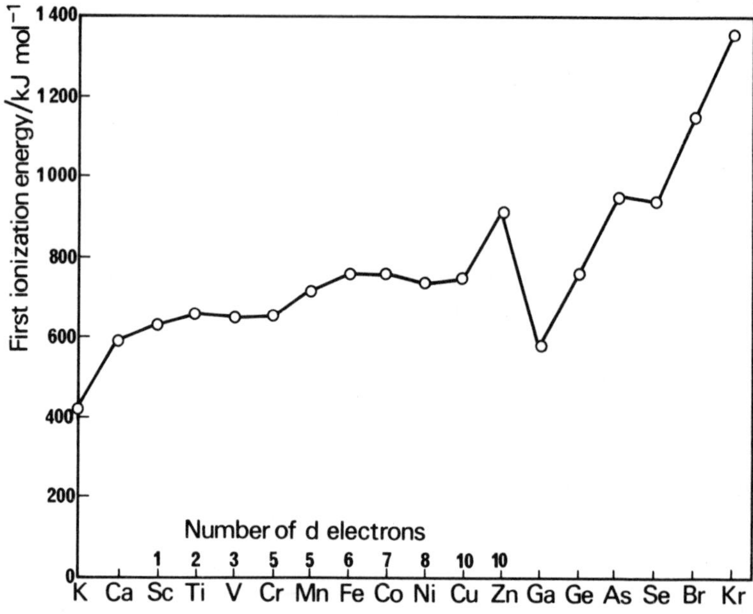

Figure 12 First ionization energies of the elements in the first long period.

Properties of chlorides and oxides

The examples given above will serve to show the periodicity of physical properties of the elements and, as was said earlier, the physical nature of the atom determines the way the atom will behave chemically—its reactivity, the type of bonds it can form, and so forth. In turn, it is possible to extend periodicity of properties to the physical nature of the compounds of a given type, for example the oxides or chlorides of the elements. It has for a long time been known that there is a parallel between the melting point of the chlorides and their electrical conductivity. The covalent chlorides (see Chapter 3) have low electrical conductivities and are volatile, that is they have relatively low enthalpies of fusion and vaporization and low melting and boiling points. The ionic chlorides have high electrical conductivities, high enthalpies of fusion, and high melting points. Figure 13, in the same arrangement as the periodic table, shows both group and period relationships in the properties of chlorides at group valency. The conductivities quoted are molar conductivities in the fused state, since it is not possible to eliminate the effect of hydrolysis in solution. A distinction is made between those chlorides which are good conductors (ionic)—enclosed in boxes—and those which are very poor conductors or non-conductors (covalent). It will be found that the

(*a*) denotes m.p./K
(*b*) denotes b.p./K
(*c*) denotes $\Lambda / \Omega^{-1}\ cm^2\ (mol\ of\ Cl)^{-1}$

	LiCl	**$BeCl_2$**							**BCl_3**	**CCl_4**	
(*a*)	883	677							166	250	
(*b*)	1 650	(825)							286	350	
(*c*)	175	0.088							0	0	
	NaCl	**$MgCl_2$**							**$AlCl_3$**	**$SiCl_4$**	**PCl_5**
(*a*)	1 081	988							456 subl.	203	421
(*b*)	1 748	1 690							433	331	–
(*c*)	141	32							1.5×10^{-5}	0	0
	KCl	**$CaCl_2$**	**$ScCl_3$**	**$TiCl_4$**			**CuCl**	**$ZnCl_2$**	**$GaCl_3$**	**$GeCl_4$**	
(*a*)	1 045	1 055	1 230	250			703	548	350	203	
(*b*)	1 680	(2 300)	1 250	410			1 960	1 005	–	357	
(*c*)	112	58	15	0			94	0.02	–	0	
	RbCl	**$SrCl_2$**	**YCl_3**	**$ZrCl_4$**	**$NbCl_5$**		**AgCl**	**$CdCl_2$**	**$InCl_3$**	**$SnCl_4$**	
(*a*)	990	1 148	970	710	480		728	841	859 subl.	240	
(*b*)	1 650	2 750	(1 780)	1 180	523		1 840	1 243	770	386	
(*c*)	86	62	9.5	0.85	2×10^{-7}		118	58.5	14.7	0	
	CsCl	**$BaCl_2$**	**$LaCl_3$**	**$HfCl_4$**	**$TaCl_5$**	**WCl_6**		**$HgCl_2$**	**$TlCl_3$**	**$PbCl_4$**	
(*a*)	918	1 233	1 143	590 subl.	484	550		550	298	260	
(*b*)	1 573	2 100	(2 000)	–	713	610		577	–	410	
(*c*)	77	71	29	–	3×10^{-7}	2×10^{-6}		3×10^{-2}	3×10^{-3}	2×10^{-5}	

Figure 13 Melting points, boiling points, and molar conductivities Λ of some chlorides of the elements at group valency. The boxes enclose the ionic chlorides.

ionic chlorides melt at or above 700 K, whilst covalent chlorides usually melt well below this temperature (varying somewhat with molecular mass).

Figure 14 illustrates one of the physical properties of oxides, the enthalpy of formation. The $\Delta H_f^{\ominus}$ values are in terms of one mole of oxygen atoms, that is 16 g of combined oxygen, since this allows comparison between oxides of various elements of different valencies. The elements with the largest negative enthalpy change when reacting with a given mass of oxygen are magnesium, aluminium, and calcium. It will be noted that the pattern in the short period is reproduced broadly in the long period, with the transition elements showing comparatively small variation in enthalpy of formation (but note the effect of variable valency). The close relationship between physical and chemical properties is exemplified by the following facts: those oxides which have positive enthalpies of formation (such as chlorine) are unstable or explosive; oxides having a small negative enthalpy of formation are very easily reduced, and may act as oxidizing agents (SO_3, CrO_3, Mn_2O_7, Cu_2O, etc.); oxides having a very large negative enthalpy of formation are very difficult to reduce (MgO, Al_2O_3, etc.).

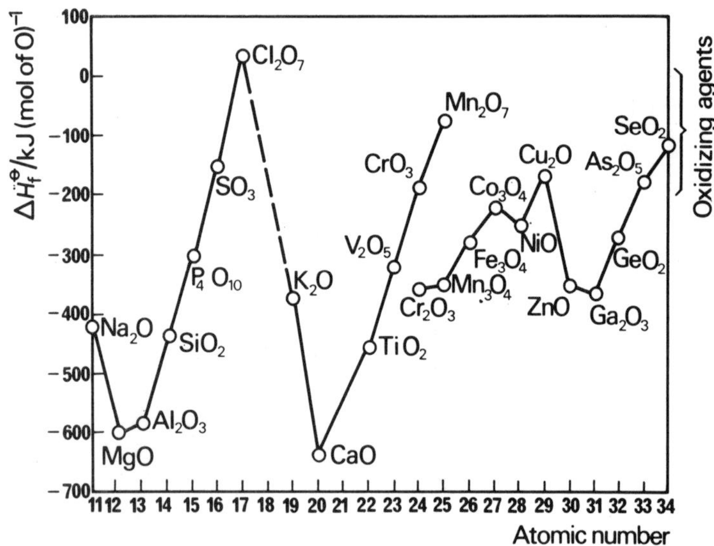

Figure 14 Enthalpies of formation of oxides.

Exercises

1 Construct a graph similar to figure 9, showing the melting points and enthalpies of fusion of elements 11 to 18 and 19 to 36.

2 Compare the extraction of aluminium from its oxide, with the reduction of the oxides of phosphorus and sulphur, in the light of figure 14.

3 Construct a graph of the number of naturally occurring isotopes of the elements against atomic number, and comment on the graph. (Figures are given in tables 5 to 13.)

Valency 3

Ionic bonding

The valency, or valence, of an element is its combining power. Valency bonds are classified into two main types, known as ionic (electrovalent) bonds and covalent bonds. An ionic bond is formed by the *transfer* of one electron from a metal atom, which thereby becomes positively charged, to a non-metal atom, which thereby becomes negatively charged:

$$\left.\begin{array}{l} M - e^- = M^+ \\ X + e^- = X^- \end{array}\right\} \text{compound } M^+X^-$$

The compound is held together by the electrostatic attraction of the oppositely charged ions. Transfer of more than one electron produces ions with more than one electric charge.

There is an energy change in each case. The energy needed to remove an electron from a metal atom gives the (first) *ionization energy* (which is calculated per mole of atoms):

$$M(g) - e^- = M^+(g)$$

(see figures 11 and 12, Chapter 2). There are various ways in which information about ionization energy can be given (see also tables 2 and 3). On the other hand, energy is given out when a halogen atom (for instance) combines with an electron:

$$X(g) + e^- = X^-(g)$$

and this gives the *electron affinity* (also calculated per mole of atoms) of the atom. Electron affinity is somewhat difficult to determine independently, but figures are available for the halogens (see table 11 in Chapter 6). If we have a metal atom and a non-metal atom, where the electron affinity of the non-metal is greater than the ionization energy of the metal, combination may be expected to take place. However, the position in a normal reaction is complicated by other changes which in turn involve energy. The metal is a solid, whose atoms must be separated; the halogen may be a gas, but energy is needed to dissociate the molecules to atoms. If the final stage is to produce a solution, then there is an energy of hydration to consider, and, in the case of a crystal, there is the lattice energy. We must consider the algebraic sum of all changes before we know from theoretical principles whether the reaction will take place. In practice, metals which have one or two electrons in the outer shell normally form ionic compounds. There are a few trivalent metals which produce trivalent ions, and a very few cases of tetravalent metal ions. The reason for this will become apparent on studying tables 2 and 3, which give one way of presenting ionization energy data for the elements (see also figures 11 and 12).

I	II	III	IV
Li 520	**Be** 1 324	**B** 2 300	**C** 3 560
Na 494	**Mg** 1 091	**Al** 1 720	**Si** 2 500

Table 2 Average ionization energies to produce group valency/kJ mol^{-1} (short periods).

K 418	**Ca** 870	**Sc** 1 426	**Ti** 2 200	**V** 3 100
Rb 401	**Sr** 802	**Y** 1 271	**Zr** 1 920	**Nb** 2 590
Cs 376	**Ba** 732	**La** 1 162	**Hf** 1 750	

Table 3 Average ionization energies to produce group valency/kJ mol^{-1} (long periods).

These tables show the average energy required to remove each of the electrons involved up to the group valency. The solid line in table 2 indicates the limit of the highly electropositive metals which normally give cations by removal of one or two electrons. The broken line divides metals from non-metals. As the required energy increases, it becomes less likely that electrons can be removed in cation formation, that is the element becomes less 'electropositive' (more 'electronegative'). Table 3 gives the data for the elements at the beginning of the long periods. The solid line again divides off the metals which normally give cations, and the broken line encloses the metals which occasionally give cations (Zr^{4+} and Hf^{4+}). It will be seen from figure 13, however, that the chlorides of these elements are not ionic. To a chemist, a metal is an element which forms cations very readily—it is 'electropositive'—and the element which does this most readily, caesium, is therefore regarded as the most typically metallic element. The change $Cs(g) - e^- = Cs^+(g)$ needs less energy than for any other element.

Similarly, the most typically non-metallic element, fluorine, is the one with the greatest tendency to form anions—it is the most 'electro-negative'. There are a number of monovalent anions, a few divalent, and no simple trivalent anions (in solution).

Covalent bonding

A *covalent bond* is formed by the *sharing* of two electrons between two atoms. Generally, these electrons come one from each atom, and when shared they contribute to the structure of each; for example

$$:\underset{\cdot\cdot}{\ddot{\mathrm{Cl}}}\cdot \;+\; \cdot\underset{\cdot\cdot}{\ddot{\mathrm{Cl}}}: \;=\; :\underset{\cdot\cdot}{\ddot{\mathrm{Cl}}}:\underset{\cdot\cdot}{\ddot{\mathrm{Cl}}}: \text{ or } \mathrm{Cl{-}Cl}$$

The shared pair of electrons is said to be in a molecular orbital, which can accommodate two electrons of opposite spin, as in the case of an atomic orbital.

In some molecules the atoms are united by multiple covalent bonds; the nitrogen molecule in the following example has three bonds with a total of six shared electrons in three molecular orbitals:

$$:\dot{\dot{\mathrm{N}}}\cdot \;+\; \cdot\dot{\dot{\mathrm{N}}}: \;=\; :\mathrm{N}:::\mathrm{N}: \text{ or } \mathrm{N{\equiv}N}$$

In special cases, a covalent bond may be formed by two electrons from the same atom, which is termed the *donor*; the atom with which they are shared is termed the *acceptor*. The oxygen atom in water and the nitrogen atom in ammonia are two examples of donors: in each case there are one or more lone (unshared) pairs of electrons on the donor atoms. Boron is often an acceptor, in its tricovalent compounds, in which it has six shared electrons. The bond, known as a *coordinate bond* (dative covalency), is shown by a line, as in the case of a covalent bond in general. However, the donor is often shown as having a positive charge (having to some extent lost electrons), and the acceptor as negative:

$$\begin{array}{ccc} \begin{array}{c} \mathrm{H} \\ \mathrm{H}:\underset{\cdot\cdot}{\ddot{\mathrm{N}}}: \\ \mathrm{H} \end{array} + \begin{array}{c} \mathrm{F} \\ \underset{\cdot\cdot}{\ddot{\mathrm{B}}}:\mathrm{F} \\ \mathrm{F} \end{array} & = & \begin{array}{c} \mathrm{H}\;\;\mathrm{F} \\ \mathrm{H}:\underset{\cdot\cdot}{\ddot{\mathrm{N}}}:\underset{\cdot\cdot}{\ddot{\mathrm{B}}}:\mathrm{F} \\ \mathrm{H}\;\;\mathrm{F} \end{array} \end{array}$$

$$\text{or}\quad \begin{array}{ccccccc} & & \mathrm{H} & & \mathrm{F} & & \\ & & | & & | & & \\ \mathrm{H} & — & \overset{+}{\mathrm{N}} & — & \overset{-}{\mathrm{B}} & — & \mathrm{F} \\ & & | & & | & & \\ & & \mathrm{H} & & \mathrm{F} & & \end{array}$$

Note that this compound is not ionic, and there is no essential difference between the coordinate bond and other covalent bonds, once formed.

$$\begin{array}{c} H \\ | \\ H{-}\ddot{N}{:} \\ | \\ H \end{array} + H^+ = \left[\begin{array}{c} H \\ | \\ H{-}N{-}H \\ | \\ H \end{array}\right]^+$$

ammonium ion, formed by coordination of ammonia with hydrogen ion

$$H_2\ddot{O}{:} + M^+ = \left[H_2\ddot{O}{-}M\right]^+$$

hydrated metal cation, formed by coordination of water with a cation (including H^+)

Note that, in these two examples, the original positive charge is shared by the whole ion.

Unlike an ionic bond, a covalent bond has direction in space with respect to the atom. This of course accounts for the tetrahedral carbon atom of organic chemistry, but it also gives particular shape to other molecules.* The ammonia molecule is not flat, but the three electron-pair bonds to hydrogen are repelled by the lone pair of electrons on the nitrogen atom so that the molecule is pyramidal, something like methane but with two electrons taking the place of one hydrogen atom. Similarly, water is not a linear molecule, but angular, or V-shaped, with an HOH bond angle of approximately 104°.

As in the case of electrovalency, there are energy changes involved in the formation of a covalent bond. Each atom must be obtained separately from its fellows, and electrons have to change orbitals. The resulting molecule has a bond energy, and the higher the bond energy, the more stable the compound. This bond is within the molecule; molecules of covalent substances, on the other hand, often have little external attraction for each other, which is another way of saying that the substances are volatile, because little kinetic energy is necessary to separate the molecules and give the gaseous state. All volatile substances are covalent. However, there are some covalent molecules, formed between polyvalent atoms, which are non-volatile because they form giant networks (silicates, for example).

Fajans's rules

Once the different types of bonds had been recognized, Fajans set out a series of three rules by which it was possible to predict whether a given bond was likely to be ionic or covalent. At the time, it was realized that most bonds were largely one or the other, and at first the compounds with bonds of intermediate character were ignored.

*See *Structure in Inorganic Chemistry* by E. Sherwin in this series.

The three rules are concerned with the size of the ions which would be produced, their charge, and their structure.

1 *Ionization is favoured by large size in the case of the metal* (*cation*), *and by small size in the case of the non-metal* (*anion*).
This is explained in the first place by the work done in removing an electron from an atom. Consider a single electron in the outer shell of an atom, radius 1 unit, and a second in an atom of radius 2 units. The force between the nucleus and the electron is given by Q_1Q_2/d^2, where Q_1 and Q_2 are the charges on nucleus and electron, and d is the distance between them. If we consider the charge on the nucleus to be one positive unit (which will be the charge on the ion), the two forces will be divided by 1^2 and 2^2 for the small and large atoms, so that the second force will then be only a quarter of the first. This argument is an over-simplification, since in fact the change in nuclear charge cannot be ignored, and there is also the screening effect of the other electrons to consider. However, measurements of the ionization energy of elements in the same group show a decrease with increasing atomic number, that is with increasing atomic radius. This applies to s-block and p-block elements but not normally to d-block elements, which do not change so much in size with increasing atomic number; in their case, there is less difference in ionization energy in successive periods. The effect of size on non-metal atoms is less obvious than for metals; here we are considering the addition of an electron to a neutral atom. When the atom is small, the electron can approach close to the nucleus when compared with a large atom. In each, the charge on the nucleus is being 'shared out' with the extra electron; this can be done more effectively in the small atom, with the closer approach.

2 *Ionization is favoured by small ionic charge, whether cation or anion.*
The alkali metals are the most highly electropositive group, and are followed by the alkaline earth metals. The difference is that the latter need to lose two electrons, whereas the former lose only one. It is easier to remove one electron—less energy is used—than to remove two. Similarly, it is easier for an atom to acquire one electron than two, which in turn is much easier than three.

3 *Ionization is favoured in those cases where the resulting ion has the electron configuration of a noble gas.*
Sodium and magnesium are examples of this—both give ions with the electron configuration of neon: 2,8. Copper, with the electron configuration 2,8,18,1, can give copper(I) ions, with the structure $2,8,18^{+}$ and also copper(II) ions, with the structure $2,8,17^{2+}$. However, both are susceptible to attack by complexing agents. Similarly, manganese gives an ion with the structure $2,8,13^{2+}$, which is readily oxidized.

Diagonal relationship

As a result of the operation of the first two rules given above, we find that there is in the periodic table a series of elements of approximately equal electronegativity which are diagonally related:

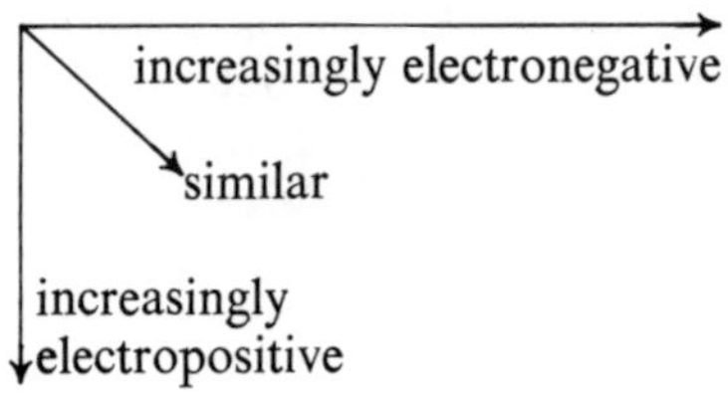

In the short periods, we find a strong resemblance between the first element in one group and the second element in the next group:

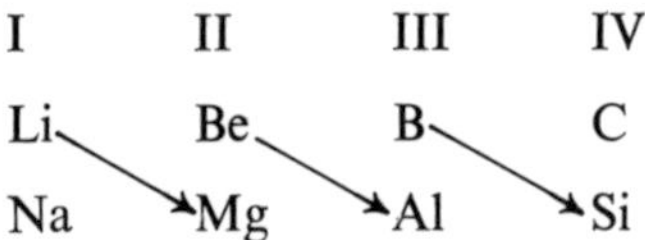

These are the three best examples of this; the student should make a comparison of the properties of one of these pairs to show how close is the resemblance. Then again, we find the division between metals and non-metals following a roughly diagonal line in the same sense.

Li	Be	B	C	N	O	F
Na	Mg	Al	Si	P	S	Cl
K	Ca	Ga	Ge	As	Se	Br
Rb	Sr	In	Sn	Sb	Te	I

In this diagram, the d-block elements have been omitted. The elements which fall near the line which has been drawn usually have properties which are neither strongly metallic nor strongly non-metallic; germanium, arsenic, and antimony are examples, and are sometimes termed *metalloids*.

Electronegativity

Electronegativity can be defined as the tendency for an atom, when combined in a molecule, to attract electrons; it has been related to the ionization energy and to the electron affinity, and indeed the mean of these two quantities has provided one scale of values. Pauling derived a scale from a consideration of bond energies, but with somewhat arbitrary assumptions. His scale of electronegativity values is broadly in agreement with other tables and has been used in plotting figure 15. This figure is an abbreviated periodic table, omitting the d-block elements, and it shows the Pauling electronegativity values of each element. The lines have been drawn by interpolation at values of 1.0, 1.5, 2.0, and so on, to link points of equal (hypothetical) electronegativity values; it will be seen that these lines are roughly diagonal in the same sense as diagonal relationships.

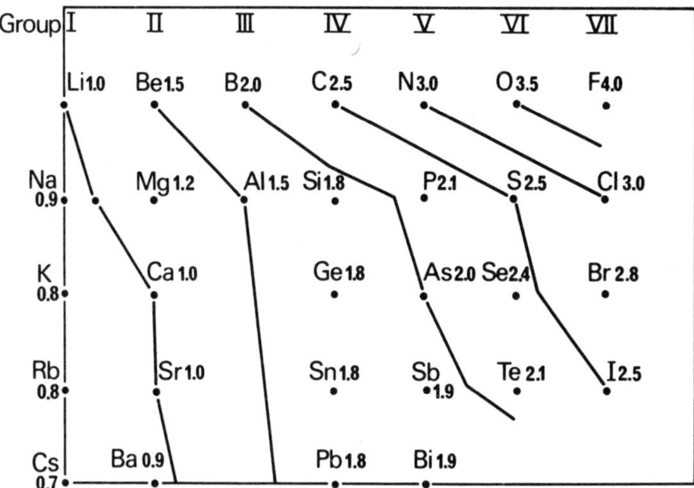

Figure 15 Electronegativity values of s- and p-block elements.

It can be said in general that the bonds formed between elements of widely differing electronegativity will be strong, and that the bonds are more likely to be ionic, the greater the difference in the values for the two elements.

Exercises

1 Show the effect of the diagonal relationship by comparing the chemistry of lithium and magnesium.

2 Repeat this exercise with boron and silicon.

3 Chart the ionic radii and ionization energies of the alkali metals *or* the alkaline earth metals.

4 Periodicity of Properties: Chemical

Examples were given in Chapter 2 of the way in which the physical properties of the elements show a periodic variation. It was also pointed out that the physical nature of the atom determines how it will behave chemically. The chemistry of the 'simple' elements—the s- and p-block elements, not the transition elements—is a function of the number of electrons in the outer shell of the atom, together with the size, or atomic radius. This is summarized in the form of Fajans's rules, which were dealt with in the previous chapter.

Short periods

In the short periods, the electrons are building up in the same shell right through a period, and elements in successive groups have the group number of electrons in the outer shell, that is the number increases by one each time. We find on traversing a period that the elements become progressively less metallic, or increasingly electronegative. This is summarized in figure 15 (see also table 2 in Chapter 3). Each period starts with an alkali metal, which is highly electropositive; all its compounds are highly ionized, with the metal always giving a monovalent cation, M^+. The metal reacts very readily with oxygen to give a highly basic oxide, and in turn this reacts with water to give a highly basic, soluble hydroxide.

The second element of the second short period is the strongly electropositive alkaline earth metal, magnesium: most of its compounds are ionic, with the metal forming a divalent cation, Mg^{2+}. The metal is readily oxidized, to give a definitely basic oxide, which reacts with water to form a very slightly soluble, strongly basic hydroxide. The second element of the first short period, beryllium, is not so strongly electropositive as magnesium and its compounds are not always ionic. However, its oxide is mainly basic.

The third element in each of the short periods is amphoteric, that is it gives an oxide which can be basic or acidic: boron is regarded as a non-metal, aluminium as a metal, and both are trivalent. The fourth and later elements in these periods are all non-metals, and their oxides are progressively more acidic up to group VII. The oxides react with water to form acids of increasing strength. Indeed, the hydrides of group VII elements are acids (in water), as well as the 'hydroxides'. The members of this group give rise to simple anions, Hal^-.

The hydrides of the elements form a series of varying properties, such that the hydrides of the alkali metals are salt-like, non-volatile solids which on fusion show the presence of *negative* hydrogen ions, H^-. This is because of the strongly electropositive nature of the metal. The middle-period elements form volatile, covalent hydrides (the vast range of hydrocarbons; silicon hydrides), while the later elements—groups VI and VII—form hydrides which, while still volatile, are ionic in aqueous solution, and give hydrogen ions, that is these hydrides are acidic. The hydrides of group VII elements are markedly more acidic than those of group VI. Some properties of hydrides are summarized in table 4 and figure 16.

I	II		III	IV	V	VI	VII
Li	Be		B	C	N	O	F
Na	Mg		Al	Si	P	S	Cl
K	Ca	Sc Ti . . . (Ni) Cu (Zn)	Ga	Ge	As	Se	Br
Rb	Sr	Y Zr Pd		Sn	Sb	Te	I
Cs	Ba	La Hf Pt	Volatile covalent hydrides				
Salt-like hydrides; react with water to give H_2		Transition metal hydrides; non-stoichiometric; react with acids to give H_2	mostly hydrolysed by water (not C)		alkaline in water	acidic in water	

Notes
The structure of Be and Mg hydrides is uncertain, but they are probably not salt-like, although they give hydrogen with water.

The transition metals do not all form definite hydrides. Those that do form are non-stoichiometric: e.g. Ti and Zr absorb H_2 to give a compound with the approximate composition MH_2, which dissolves or alloys with an excess of metal in varying amounts. Pd has a very great power of dissolving H_2, and a heated Pd foil is 'transparent' to H_2. Pt has a similar power to a lesser extent.

The periodic variation of the bond energies of the covalent hydrides is illustrated in figure 16; the boiling points give similar relationships.

Table 4 The hydrides of the elements.

The hydrides of group V elements are worthy of special mention: NH_3 and PH_3 both give solutions which are somewhat basic. This is because both these molecules have a lone (unshared) pair of electrons not being used in the covalent bonds between nitrogen or phosphorus and hydrogen atoms. In solution, therefore, the unshared pair of electrons makes the compound behave as a donor, so that coordination can take place with H^+ ions ($H_3N—H^+$) to give NH_4^+ and leave an excess of OH^- ions.

It is possible to make comparisons of several other compounds, such as chlorides, in this way. However, given compounds are not always formed right through a period, and sometimes there is a rather less obvious variation in properties than in the case of the oxides and hydrides quoted. All the variation in properties shown is compatible with the

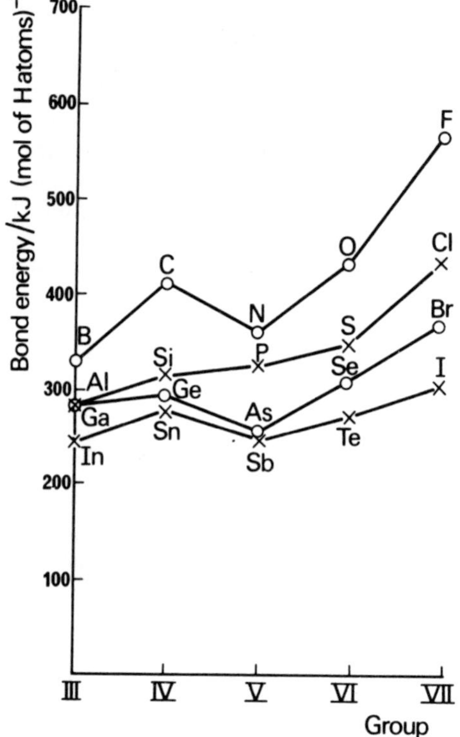

Figure 16 Bond energies of covalent hydrides.

statement already made, that the elements along a period become progressively more electronegative. The electronegativity values already charted (figure 15) summarize chemical properties to some extent. Elements of widely differing electronegativity react readily, if not violently, usually to give compounds which are strongly ionized.

Long periods

The first long period starts in the same way as the short periods, with an alkali metal and an alkaline earth metal, showing a similar relationship to that between sodium and magnesium. There is then a number of metals without parallel in the short periods, known as transition metals, whose properties vary more slowly than other consecutive elements. These metals are followed by p-block elements which continue the pattern of the short periods, and each is broadly similar to the elements of the same group above them in the table. The group VII elements (the halogens) were mentioned above, and the long periods again have halogens in the same group. Both short and long periods finish with a group VIII element, a gas at one time called 'inert'. These gases are indeed unreactive, and each one has a 'completed' shell of eight electrons as its outer shell. (The shell of eight electrons is known as an octet, and is a stable structure as an outer shell of an atom.) However, the gases do form some compounds, and the name 'noble gases' is now preferred. The discovery of this group only began in 1895, but they were fitted in to the edge of the table without difficulty, and their discovery stimulated the theory of the electronic structure of the elements. It was soon realized that other atoms often react in such a way that their electronic structure approaches that of an 'inert' gas: those atoms which have six or seven electrons in the outer shell acquire two or one more, while those atoms which have only one or two electrons in the outer shell tend to lose them. This approach stimulated the development of valency theory, as outlined in Chapter 3.

Variation of properties in the transition metals

The elements in parentheses, as it were, show a slow progression of properties. They have a number of interesting properties which are not found all at once among other metals, and therefore they are usually treated separately.

These metals appear in the d-block of elements, since in their atoms the d sub-shell is filling up. Figure 4 in Chapter 1 gives the electron configurations for the transition metals of the first long period; it will be seen that, in general, the metals have two electrons in the 4s orbital, while the 3d orbital has one extra electron in successive elements. The electrons in the outer shell are readily available and can be relatively easily removed, so the metals usually form cations, such as Mn^{2+}, in a stable range of compounds. The electronic structure of these cations does not correspond to a noble gas configuration, and in accordance with Fajans's rules they are not as stable as a cation such as Mg^{2+}. The electrons of the 3d orbitals can be attacked by oxidizing agents, which thereby compel the transition metals to exert a higher valency, up to the group number as maximum, and this high valency must be a covalency.

The transition elements (which are treated in more detail in Chapter 7) form a series of elements with considerable similarity: their properties are changing more slowly than those of neighbouring elements in the short periods (or those of elements in later groups of the long periods). Their atoms are similar in size as well as in outer electron configuration.

The f-block elements are known as *inner transition elements*; in their case extra electrons are being added in 4f or 5f orbitals, when there are electrons present in *two* further shells beyond the one which is growing. This makes neighbours more closely similar even than neighbours of the d-block transition series.

The various long periods show periodicity in the occurrence, and placing, of the transition metals; the solution of the problem of their electronic structure provided extra evidence of the usefulness of electronic theory applied to the periodic table.

Exercises

1 List the chlorides of the elements in the first two short periods (Li to F and Na to Cl). Show their structures and reaction with water, if any.

2 Compare the oxides of beryllium, boron, aluminium, and silicon as examples of the diagonal relationship.

5 The s-Block Elements: Groups I and II

This block contains only metals, and these are highly electropositive. Group I elements are known as the alkali metals, and group II elements as the alkaline earth metals; beryllium alone (the first element of group II) shows any departure from the purely basic behaviour normally found in the oxides and hydroxides of these metals. Beryllium, indeed, shows a number of differences in behaviour from the other metals in group II, and in many ways it resembles aluminium, the second element in group III. This is one of the examples of the diagonal relationship already mentioned in Chapter 3. The following remarks should therefore be taken to exclude beryllium, in most cases.

Tables 5 and 6 give the physical constants for the elements, and figures 17 and 18 illustrate the variation in the size of the atoms and ions, and in the ionization energy. It will be noted that the ionic and covalent radii increase steadily in both groups as the atomic number increases, that the group I radii are greater than those of the corresponding elements in group II, and that the ionic radii are always considerably less than the covalent radii. Figure 18 compares the first ionization energies for group I metals, and the first and second ionization energies for group II metals. It will be seen that the ionization energies for group I are low and decrease with increasing size of the atoms, and for group II they are higher, but fall more rapidly as the covalent radius increases. The reason for the exceptional behaviour of beryllium is found in its small size and high ionization energy, compared with the other metals. Reference to tables 5 and 6 will show that the melting points, enthalpies of fusion, boiling points, and enthalpies of vaporization all fall progressively with increasing size of the atoms, and these properties like the ionization energies are inversely dependent on the atomic size within a group.

Table 5 Properties of the group I elements (alkali metals).

	Li	Na	K	Rb	Cs
Atomic number	3	11	19	37	55
Electron configuration	(2)2s^1	(2,8)3s^1	(2,8,8)4s^1	(2,8,18,8)5s^1	(2,8,18,18,8)6s^1
Relative atomic mass, A_r	6.941	22.989 8	39.102	85.467 8	132.905 5
Melting point/K	454	371	336	312	302
Enthalpy of fusion/kJ mol^{-1}	3.01	2.59	2.30	2.18	2.09
Boiling point/K	1 604	1 163	1 040	975	960
Enthalpy of vaporization/kJ mol^{-1}	133	90	77.5	69.1	65.9
First ionization energy/kJ mol^{-1}	519	498	418	401	376
Covalent radius/pm	134	154	196	211	225
Ionic radius/pm (M^+)	60	95	133	148	169
Standard electrode potential, $E^\ominus$/V (M^+/M)	−3.02	−2.71	−2.92	−2.99	−3.02
Enthalpy of hydration of M^+ion/kJ mol^{-1}	−519	−407	−322	−301	−276
Molar conductivity of M^+ ion /Ω^{-1} cm^2 mol^{-1}	38.7	50.1	73.5	77.8	77.3
Number of naturally occurring isotopes	2	1	3	2	1

Table 6 Properties of the group II elements (alkaline earth metals).

	Be	Mg	Ca	Sr	Ba
Atomic number	4	12	20	38	56
Electron configuration	(2)2s^2	(2,8)3s^2	(2,8,8)4s^2	(2,8,18,8)5s^2	(2,8,18,18,8)6s^2
Relative atomic mass, A_r	9.012 18	24.305	40.08	87.62	137.34
Melting point/K	1 553	924	1 124	1 073	1 123
Enthalpy of fusion/kJ mol^{-1}	11.6	9.0	8.8	9.2	7.7
Boiling point/K	3 040	1 380	1 710	1 650	1 910
Enthalpy of vaporization/kJ mol^{-1}	293	129	150	139	151
First ionization energy/kJ mol^{-1}	900	740	590	548	502
Second ionization energy/kJ mol^{-1}	1 760	1 450	1 150	1 061	836
Covalent radius/pm	90	130	174	192	198
Ionic radius/pm (M^{2+})	31	65	99	113	135
Standard electrode potential, $E^\ominus$/V (M^{2+}/M)	−1.70	−2.34	−2.87	−2.89	−2.90
Enthalpy of hydration of M^{2+} ion/kJ mol^{-1}	−2 981	−2 082	−1 760	−1 600	−1 450
Molar conductivity of M^{2+} ion/Ω^{-1} cm^2 mol^{-1}	90.0	106.1	119.0	118.9	127.2
Number of naturally occurring isotopes	1	3	6	4	7

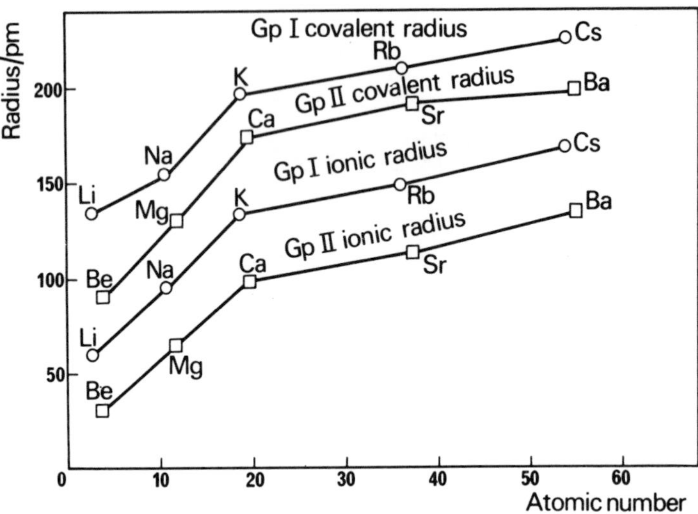

Figure 17 Covalent and ionic radii of the s-block elements.

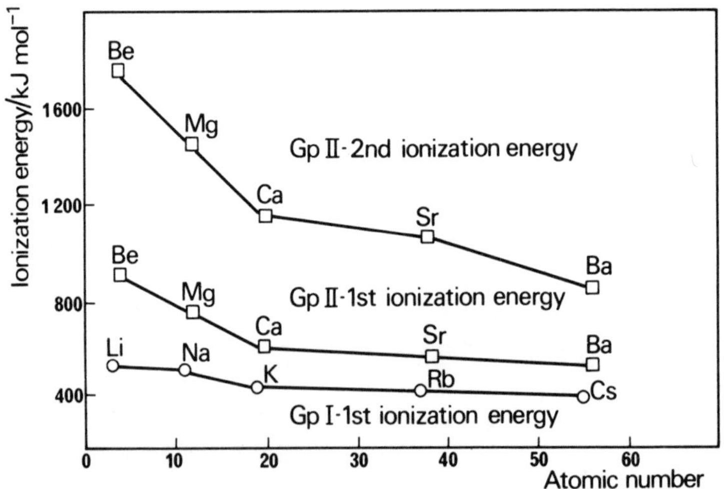

Figure 18 Ionization energies of the s-block elements.

The metals react readily with cold water, except for magnesium, which reacts with hot water. The hydroxides are soluble and strongly basic, though again magnesium is an exception with a hydroxide which is only very slightly soluble but still definitely basic.

The chemistry of the alkali metals is dominated by their strong tendency to form ionic compounds. The extent of hydration of the cations M^+ decreases as the size of the ion increases; lithium compounds are usually hydrated, and the lithium ion has a comparatively low conductivity in solution. The exceptionally high enthalpy of hydration of Li^+ explains why the standard electrode potential for Li^+/Li is 'out of order' when compared with the other alkali metals, and has a value equal to that for Cs^+/Cs. Sodium compounds are usually hydrated, although more often when polyvalent anions are involved—NaCl has no water of crystallization, whilst Na_2CO_3, Na_2SO_4, and Na_3PO_4 all have. Potassium compounds are not usually hydrated, and K^+ is much more 'mobile' than Li^+.

The chemistry of the alkaline earth metals broadly follows that of the alkali metals, but the tendency to form ionic compounds is not quite so great. Magnesium, whilst not being so exceptional as beryllium, does form a number of covalent compounds (including the well-known Grignard reagents), and it does not fall quite in place when considering the group relationship: barium, strontium, calcium, magnesium. This is the order of *decreasing* reactivity, of *decreasing* tendency to ionize, and of *increasing* tendency for the ion M^{2+} to be hydrated. (Compare, for example, $BaCl_2 \cdot 2H_2O$ and $CaCl_2 \cdot 6H_2O$; $BaSO_4$ and $CaSO_4 \cdot 2H_2O$.) In this group, the compounds such as $M^{2+}SO_4^{2-}$, where the charges on cation and anion are equal, have less water of crystallization than compounds where the charges are unequal.

Exercises

1 List the number of molecules of water of crystallization in the more common sodium and potassium compounds. Hence determine the number of hydrated compounds, and the average extent of hydration in these compounds.

2 Explain the difference between sodium and potassium compounds, in respect of the extent of hydration of crystals, in terms of Fajans's rules.

3 Potassium manganate(VII) (potassium permanganate) and potassium dichromate(VI) (potassium dichromate) are used in quantitative analysis in place of the (cheaper) sodium compounds. Explain why this is so.

4 Make charts showing the relationship between atomic number and (*a*) enthalpy of fusion and melting point, (*b*) enthalpy of vaporization and boiling point, for the group I elements and the group II elements.

5 Repeat exercise 1 for the metals calcium, strontium, and barium.

The p-Block Elements: Groups III to VIII 6

The p-block elements vary widely in their properties, so that it is impossible to make any general statements. In the successive groups, the p electron sub-shell is building up from one to six, as shown in the summary tables. The maximum valency employs the s electrons as well as the p electrons, and is equal to the group number. However, there are often alternative valencies, which should be looked at in a group context.

Group III (table 7)

This group starts with a non-metal, boron, and the elements become increasingly metallic as the atomic number rises. Boron shows considerable similarity to silicon, and aluminium to beryllium.

Both boron and aluminium give tricovalent compounds which, having only six shared electrons, behave as if they are electron-deficient. Some of the compounds dimerize and some act as electron-acceptors and form coordination compounds.

Aluminium is a rather weakly electropositive metal, which only rarely gives ionic compounds (AlF_3 is ionic, Al_2Cl_6 is covalent). The remaining three metals have an alternative behaviour available, due to the 'inert pair' effect, normally confined to the later period elements and more prominent in later groups. In these elements, the two s electrons need not take part in valency formation, but remain inert. The valency thus becomes two less than the group number, and in this group we find certain monovalent compounds. These are very important with thallium, but decrease in importance with indium and gallium. Monovalent thallium is strongly electropositive (in accordance with Fajans's rules) forming ionic compounds which do not readily oxidize; monovalent indium is easily oxidized, and monovalent gallium is only found in a few cases, so we must conclude that the inert pair effect is more favoured, the later the period.

Group IV (table 8)

In this group we start with two non-metals, carbon and silicon. The bond between two carbon atoms is unusually strong and is not weakened by repetition, so that there is no limit to the number of carbon atoms per molecule. Carbon atoms can be arranged in unbranched or branched chains and rings, a variety of different atoms can be bonded to carbon,

	B	**Al**	**Ga**	**In**	**Tl**
Atomic number	5	13	31	49	81
Electron configuration	$(2)2s^22p^1$	$(2,8)3s^23p^1$	$(2,8,18)4s^24p^1$	$(2,8,18,18)5s^25p^1$	$(2,8,18,32,18)6s^26p^1$
Relative atomic mass, A_r	10.81	26.981 5	69.72	114.82	204.37
Melting point/K	2 570	933	303	430	720
Enthalpy of fusion/kJ mol^{-1}	22.2	10.7	5.58	3.26	4.28
Boiling point/K	2 820	2 773	2 340	2 370	2 660
Enthalpy of vaporization/kJ mol^{-1}	539	293	257	226	162
First ionization energy/kJ mol^{-1}	800	577	579	558	588
Third ionization energy/kJ mol^{-1}	3 650	2 738	2 950	2 689	2 863
Covalent radius/pm	80	125	125	150	155
Ionic radius/pm (M^{3+})	—	50	62	81	95
Standard electrode potential, $E^{\ominus}$/V (M^{3+}/M)	−0.73	−1.67	−0.52	−0.34	+0.72
Number of naturally occurring isotopes	2	1	2	2	2

Table 7 Properties of the group III elements.

	C	**Si**	**Ge**	**Sn**	**Pb**
Atomic number	6	14	32	50	82
Electron configuration	$(2)2s^22p^2$	$(2,8)3s^23p^2$	$(2,8,18)4s^24p^2$	$(2,8,18,18)5s^25p^2$	$(2,8,18,32,18)6s^26p^2$
Relative atomic mass, A_r	12.011	28.086	72.59	118.69	207.2
Melting point/K	3 770	1 693	1 230	505	600
Enthalpy of fusion/kJ mol^{-1}	—	46	32	7.2	5.1
Boiling point/K	5 070	2 970	2 970	2 630	2 030
Enthalpy of vaporization/kJ mol^{-1}	711	440	334	290	179
First ionization energy/kJ mol^{-1}	1 086	785	780	708	715
Second ionization energy/kJ mol^{-1}	2 350	1 580	1 534	1 410	1 450
Fourth ionization energy/kJ mol^{-1}	6 210	4 350	4 400	3 925	4 040
Covalent radius/pm	77	117	122	141	150
Ionic radius/pm (M^{2+})	—	—	93	112	120
Number of naturally occurring isotopes	2	3	5	10	4

Table 8 Properties of the group IV elements.

and the directed covalent bonds give rise to many cases of isomerism; thus we have vast numbers of compounds of carbon, which has made it convenient to study these compounds within the branch of chemistry known as organic chemistry.

Silicon, the second most common element on earth, is one of the elements that can be bonded to carbon, although the Si—Si bond is not as strong as the C—C bond, and there is a growing organic chemistry of the various families of carbon compounds containing silicon in addition. Silicates, because of the polyvalency of silicon, have high molecular masses in many cases, and complex structures are often found. Such compounds are non-volatile and resistant to heat (mica, asbestos, etc.); they lend themselves to the development of materials having interesting and useful properties. Germanium is a metalloid, and the remaining two elements are metals, so that there is the usual pattern of increasing electropositive behaviour as the atomic number rises. The inert pair effect is well established with tin and lead, giving rise to quite strongly ionic compounds of Sn^{2+} and Pb^{2+}. There is a difference between the two, in that two is the normal valency for lead but Sn^{2+} is a reducing agent. Germanium shows inert pair behaviour, but divalent germanium is not common. Again, therefore, this behaviour is favoured increasingly in the later periods.

Group V (table 9)

This group shows a similar pattern to group IV, with two non-metals, arsenic a metalloid but less metallic than germanium, and two metals showing definite inert pair behaviour.

Nitrogen is a 'permanent' gas which is relatively unreactive. Its trivalent compounds have an unshared pair of electrons, and such compounds readily act as donors in forming coordination compounds. Phosphorus is a volatile solid which is much more reactive than nitrogen towards oxidation. Phosphorus is not restricted to a shared octet (as nitrogen is), and can therefore exert a covalency of five. Phosphates have some similarity to silicates in the variety of their structure: there is a particular capacity to form complex acids and their salts, involving the acidic oxides of various transition metals. Arsenic, antimony, and bismuth are increasingly metallic in their behaviour. All three readily form trivalent compounds in which the metal has an inert pair, and trivalent arsenic is a reducing agent. Trivalent antimony is not markedly reducing (although it can be oxidized) and three is the normal valency for bismuth. There is, therefore, a tendency for the inert pair effect to start with an earlier period as the group number increases.

Group VI (table 10)

In this group metallic behaviour is not marked until the fourth element,

	N	**P**	**As**	**Sb**	**Bi**
Atomic number	7	15	33	51	83
Electron configuration	$(2)2s^22p^3$	$(2,8)3s^23p^3$	$(2,8,18)4s^24p^3$	$(2,8,18,18)5s^25p^3$	$(2.8,18,32,18)6s^26p^3$
Relative atomic mass, A_r	14.006 7	30.973 8	74.921 6	121.75	208.980 6
Melting point/K	63	317	—	903	543
Enthalpy of fusion/kJ mol^{-1}	0.36	0.63	27.6	20	11
Boiling point/K	77	554	889(subl.)	1 700	1 700
Enthalpy of vaporization/kJ mol^{-1}	2.8	12.5	144	58	151
Covalent radius/pm	75	106	119	138	146
Ionic radius/pm (M^{3+})	—	—	69	90	120
Standard electrode potential, $E^{\ominus}$/V (M^{3+}/M)	—	—	+0.25	+0.21	+0.32
Number of naturally occurring isotopes	2	1	1	2	1

Table 9 Properties of the group V elements.

	O	**S**	**Se**	**Te**
Atomic number	8	16	34	52
Electron configuration	$(2)2s^22p^4$	$(2,8)3s^23p^4$	$(2,8,18)4s^24p^4$	$(2,8,18,18)5s^25p^4$
Relative atomic mass, A_r	15.999 4	32.06	78.96	127.60
Melting point/K	54	387	490	723
Enthalpy of fusion/kJ mol^{-1}	0.22	1.4	5.2	18
Boiling point/K	90	718	960	1 660
Enthalpy of vaporization/kJ mol^{-1}	3.4	9.8	26.4	51
Covalent radius/pm	74	102	116	135
Ionic radius/pm (X^{2-})	140	184	198	221
Ionic radius/pm (X^{4+})	—	—	—	89
Number of naturally occurring isotopes	3	4	6	8

Table 10 Properties of the group VI elements.

tellurium. Oxygen, the most common element on earth, is again a 'permanent' gas, but more reactive than nitrogen. Most elements combine directly with oxygen, although in most cases the rate of combination is very slow at room temperature. Many metals acquire a protective film of oxide which resists further penetration unless mechanically damaged. Oxygen shows a variety of properties, because it can become an anion O^{2-} (which however cannot exist in solution as it immediately gives OH^-), or it can form two covalencies. In this state, there are two unshared pairs of electrons, and oxygen is a powerful donor. Sulphur has some similarities to oxygen, for example its readiness to combine with metals, but it is more varied in its behaviour. Again it can be ionic; S^{2-} can exist in water, though it is largely hydrolysed to SH^-. In the dicovalent state it is much less powerful a donor than oxygen. However, sulphur can expand its octet (like phosphorus but not oxygen), and therefore show higher covalencies, up to six. Sulphur forms a number of oxoacids, somewhat similar to the phosphorus oxoacids, but without the power to form complex heteropolyacids with transition metal oxides.

Selenium has some similarities to sulphur, but the range of properties is less wide, and there is less variety of acids; selenium dioxide is only weakly acidic, and selenium has some metalloid characteristics. Tellurium dioxide is even less acidic than selenium dioxide, and tellurium is a little more metallic in its properties than selenium although it is still classified as a non-metal. There is some sign of inert pair behaviour with a small concentration of Te^{4+} in certain compounds. The last member of the group, polonium, is highly radioactive and very scarce in occurrence; in properties it shows the expected differences from tellurium, and is still more metallic in character.

Group VII—the halogens (table 11)

These elements form a very well-marked family with a gradation of properties. All are clearly non-metals, with iodine the least electronegative (astatine, a very scarce radioelement fits into this gradation). All are volatile and form acidic hydrides. However, fluorine shows certain individualities: for instance, its acid is highly associated by hydrogen-bonding, and is therefore less volatile and less strongly ionized than expected by comparison with the others. Fluorides are often different from other halides in their solubilities.

Fluorine is the most electronegative element, and compels elements to exert high valencies. It combines directly with most elements, and metal fluorides are usually more ionic than other halides. Many noble gas compounds contain fluorine. Chlorine and the other halogens follow in order after fluorine, being highly electronegative and very reactive with metals, often in the cold (compare oxygen, which usually reacts readily only on heating). Fluorine is very reluctant to combine with

	F	Cl	Br	I
Atomic number	9	17	35	53
Electron configuration	(2)$2s^22p^5$	(2,8)$3s^23p^5$	(2,8,18)$4s^24p^5$	(2,8,18,18)$5s^25p^5$
Relative atomic mass, A_r	18.998 4	35.453	79.904	126.904 5
Melting point/K	40	171	266	386
Enthalpy of fusion/kJ mol^{-1}	0.25	3.2	5.2	7.8
Boiling point/K	85	238	332	453
Enthalpy of vaporization/kJ mol^{-1}	3.3	10	15	21
Electron affinity/kJ mol^{-1}	335	355	332	301
Covalent radius/pm	72	99	114	133
Ionic radius/pm (X^-)	136	181	195	216
Enthalpy of hydration of X^- ion/kJ mol^{-1}	401	279	243	201
Molar conductivity of X^- ion/Ω^{-1} cm^2 mol^{-1}	55.4	76.4	78.3	76.8
Number of naturally occurring isotopes	1	2	2	1

Table 11 Properties of the group VII elements (the halogens).

	He	Ne	Ar	Kr	Xe
Atomic number	2	10	18	36	54
Electron configuration	$1s^2$	(2)$2s^22p^6$	(2,8)$3s^23p^6$	(2,8,18)$4s^24p^6$	(2,8,18,18)$5s^25p^6$
Relative atomic mass, A_r	4.002 60	20.179	39.948	83.80	131.30
Melting point/K	0.9	24	84	116	161
Enthalpy of fusion/kJ mol^{-1}	0.01	0.32	1.1	1.5	2.1
Boiling point/K	4	27	87	120	166
Enthalpy of vaporization/kJ mol^{-1}	0.08	1.8	6.3	5.5	13.6
First ionization energy/kJ mol^{-1}	2 639	2 078	1 519	1 349	1 169
Atomic radius/pm	93	112	154	169	190
Number of naturally occurring isotopes	2	3	3	6	9

Table 12 Properties of the group VIII elements (the noble gases).

oxygen, and the other halogens are not highly reactive towards oxygen. However, while there are oxoacids of chlorine, bromine, and iodine, there is no oxoacid of fluorine. The halogen atoms in their monovalent compounds have three unshared pairs of electrons, but fluorine alone cannot use these electrons as a donor in forming coordinate links with oxygen because suitable orbitals are not available in fluorine.

Group VIII—the noble gases (table 12)

These elements were only discovered just before 1900, and were at first called rare or inert gases. Argon is in fact present in the atmosphere (more than one per cent, by volume, while the others are present in traces), and it is now known that compounds can be formed, particularly of krypton and xenon. Hence the preference for the title noble gases, by comparison with the noble metals.

The elements all have a complete octet, a very stable structure, and this is the reason for calling them group VIII. Until recently no normal compounds of these gases had been made although clathrate compounds had been made—'cage' molecules with an atom of argon inside the cage, but without a valency bond. Physical measurements showed the expected decrease of ionization energy on going from helium to xenon, and it was noted that the first ionization energy for xenon was less than that for some elements in other groups of the periodic table. It was therefore logical to try the effect of strongly oxidizing conditions on xenon, and this was the first of these gases to give compounds with fluorine, with or without a metal in the molecule. Perhaps the most important theoretical aspect of these gases was to aid the development of the electronic theory of valency, for it was realized that many atoms react in such a way that their electronic structure approaches that of the nearest 'inert' gas (as they were then called), that is approaches the complete octet.

Exercises

1 Trace the variation of properties of the elements from boron to fluorine, by referring to compounds formed with hydrogen, oxygen, and chlorine.

2 Repeat exercise 1 with the elements from aluminium to chlorine.

3 Show the effect of inert pair behaviour in tin and lead, by comparing briefly their divalent and tetravalent compounds.

4 Illustrate the family relationships of the halogens by reference to their compounds with hydrogen, oxygen, sodium, and carbon.

5 Tabulate briefly the properties of chlorine, argon, and potassium. What is the relationship between Cl^-, Ar, and K^+?

The d- and f-Block Elements 7

The d-block elements (transition elements)

The elements of the d-block are adding d electrons in the penultimate shell, and have from one to ten such electrons. The fact that an inner shell is changing means that the successive elements do not vary in properties in the same way that successive members of the p-block vary, where the outer shell is gaining electrons. Table 13 summarizes some physical properties of the d-block elements, and figure 19 illustrates their electron configurations. The pattern in the three series is not identical, although there are broad similarities. It should be emphasized that energy levels associated with s, p, and d orbitals are getting closer together as the principal quantum number (shell number) increases. This allows electron transitions between orbitals to take place more readily in later periods, and explains a number of points of transition metal chemistry.

Variable valency

Atoms of these elements can lose the s electrons to give ions such as Mn^{2+} and Fe^{2+}; however, the electronic structure of such ions (2,8,13 for Mn^{2+} and 2,8,14 for Fe^{2+}) does not correspond to the configuration of a noble gas, and therefore it is not specially stable. Under oxidizing conditions, the d electrons become available; the increased energy required to remove further electrons completely (which would lead to the formation of polyvalent cations) is not normally available, and the higher valencies are covalencies. Different oxidizing agents can lead to different values of these valencies—the group maximum is not always reached. One of the features of transition metal chemistry which has long been recognized is this variable valency.

Complex ions; oxidation state

The metals are also particularly prone to forming coordination complexes, such as $[Fe(CN)_6]^{3-}$ and $[Fe(CN)_6]^{4-}$; if coordination takes place with an anion, as in this case, the metal becomes part of a complex anion. The higher oxides of transition metals are indeed acidic, such as Mn_2O_7 which gives rise to manganate(VII) compounds containing the MnO_4^- ion. Because these metals frequently form complex compounds, the term 'oxidation state' is preferred to 'valency', particularly in such circumstances. The algebraic sum of the oxidation states in a compound

	Sc	**Y**	**La**	**Ti**	**Zr**	**Hf**
Relative atomic mass, A_r	44.955 9	88.905 9	138.905 5	47.90	91.22	178.49
Covalent radius/pm	144	162	169	132	145	144
Ionic radius/pm	81(3+)	90(3+)	106(3+)	68(4+)	80(4+)	81(4+)
Melting point/K	1 670	1 770	1 200	2 000	2 130	2 470
Enthalpy of fusion/kJ mol^{-1}	15.9	17.2	11.4	15.6	16.7	21.8
Boiling point/K	2 750	3 500	3 640	3 550	4 650	5 550
Enthalpy of vaporization/kJ mol^{-1}	304	393	399	428	581	671
Number of naturally occurring isotopes	1	1	2	5	5	6

	V	**Nb**	**Ta**	**Cr**	**Mo**	**W**	**Mn**	**Tc**	**Re**
Relative atomic mass, A_r	50.941 4	92.906 4	180.947 9	51.996	95.94	183.85	54.938 0	98.906 2	186.2
Covalent radius/pm	122	134	134	117	129	130	117	—	128
Melting point/K	2 140	2 680	3 120	2 200	2 900	3 650	1 530	2 420	3 440
Enthalpy of fusion/kJ mol^{-1}	17.6	26.8	31.4	13.8	27.6	35.2	16.7	23.0	33.2
Boiling point/K	3 650	5 200	5 670	2 920	5 100	5 800	2 310	3 800	5 900
Enthalpy of vaporization/kJ mol^{-1}	458	697	752	348	593	798	219	677	707
Number of naturally occurring isotopes	1	1	2	4	7	5	1	0	2

Table 13 Properties of the d-block elements (transition elements).

Relative atomic mass, A_r	55.847	58.933 2	58.71	101.07	102.905 5	106.4	190.2	192.22	195.09
Covalent radius/pm	116	116	115	124	125	128	126	126	129
Ionic radius/pm (M^{2+})	76	78	78	—	86	50	—	—	52
Melting point/K	1 810	1 750	1 730	2 270	2 240	1 830	2 970	2 720	2 040
Enthalpy of fusion/kJ mol^{-1}	15.5	15.4	17.6	25.4	22.0	16.8	29.4	26.4	19.7
Boiling point/K	3 160	4 150	3 110	4 000	4 000	3 400	4 500	4 400	4 100
Enthalpy of vaporization/kJ mol^{-1}	351	384	372	567	494	376	627	561	510
First ionization energy/kJ mol^{-1}	756	760	735	724	745	803	841	887	866
Second ionization energy/kJ mol^{-1}	1 564	1 676	1 756	1 620	1 740	1 870	1 630	1 550	1 870
Number of naturally occurring isotopes	4	1	5	7	1	6	7	2	6

	Cu	**Ag**	**Au**	**Zn**	**Cd**	**Hg**
Relative atomic mass, A_r	63.546	107.868	196.966 5	65.37	112.40	200.59
Covalent radius/pm	117	134	134	125	141	144
Ionic radius/pm	96 (1+)	126 (1+)	137 (1+)	74 (2+)	97 (2+)	110 (2+)
Melting point/K	1 360	1 230	1 335	692	594	234
Enthalpy of fusion/kJ mol^{-1}	13.0	11.4	12.8	7.4	6.2	2.3
Boiling point/K	2 855	2 450	2 980	1 180	1 040	630
Enthalpy of vaporization/kJ mol^{-1}	304	255	325	115	100	57
First ionization energy/kJ mol^{-1}	744	732	890	903	857	1 004
Second ionization energy/kJ mol^{-1}	1 952	2 115	1 935	1 730	1 630	1 805
Number of naturally occurring isotopes	2	2	1	5	8	7

Table 13 Properties of the d-block elements (transition elements)—*continued*.

is zero, and, in the two complex cyanides quoted above, the oxidation states are:

$$\begin{array}{cc} K_3\ Fe\ (CN)_6 & K_4\ Fe\ (CN)_6 \\ (+1)\times 3,\ +3,\ (-1)\times 6 & (+1)\times 4,\ +2,\ (-1)\times 6 \end{array}$$

Fe^{III} indicates the oxidation state of iron in the first compound, Fe^{II} in the second. In the case of manganese, Mn_2O_7 has $(+7) \times 2$ for manganese and $(-2) \times 7$ for oxygen; the manganate(VII) (permanganate) ion MnO_4^- has $+7$ for manganese, $(-2) \times 4$ for oxygen, and $+1$ supplied by the cation. Mn^{VII} indicates the oxidation state of manganese in both cases.

Atomic number	Element	3s	3p	3d					4s	
20	calcium	2	6						↑↓	
21	scandium	2	6	↑					↑↓	first transition series (21–29)
22	titanium	2	6	↑	↑				↑↓	
23	vanadium	2	6	↑	↑	↑			↑↓	
24	chromium	2	6	↑	↑	↑	↑	↑	↑	
25	manganese	2	6	↑	↑	↑	↑	↑	↑↓	
26	iron	2	6	↑↓	↑	↑	↑	↑	↑↓	
27	cobalt	2	6	↑↓	↑↓	↑	↑	↑	↑↓	
28	nickel	2	6	↑↓	↑↓	↑↓	↑	↑	↑↓	
29	copper	2	6	↑↓	↑↓	↑↓	↑↓	↑↓	↑	
30	zinc	2	6	↑↓	↑↓	↑↓	↑↓	↑↓	↑↓	

Atomic number	Element	4s	4p	4d					5s	
38	strontium	2	6						↑↓	
39	yttrium	2	6	↑					↑↓	second transition series (39–47)
40	zirconium	2	6	↑	↑				↑↓	
41	niobium	2	6	↑	↑	↑	↑		↑	
42	molybdenum	2	6	↑	↑	↑	↑	↑	↑	
43	technetium	2	6	↑↓	↑	↑	↑	↑	↑	
44	ruthenium	2	6	↑↓	↑↓	↑	↑	↑	↑	
45	rhodium	2	6	↑↓	↑↓	↑↓	↑	↑	↑	
46	palladium	2	6	↑↓	↑↓	↑↓	↑↓	↑↓		
47	silver	2	6	↑↓	↑↓	↑↓	↑↓	↑↓	↑	
48	cadmium	2	6	↑↓	↑↓	↑↓	↑↓	↑↓	↑↓	

Figure 19 Electron configurations of d-block elements.

		4s	4p	4d	4f	5s	5p	5d					6s	
56	barium	2	6	10		2	6						↑↓	
57	lanthanum	2	6	10		2	6	↑					↑↓	
[58–71	1st f-block]													
72	hafnium	2	6	10	14	2	6	↑	↑				↑↓	
73	tantalum	2	6	10	14	2	6	↑	↑	↑			↑↓	
74	tungsten	2	6	10	14	2	6	↑	↑	↑	↑		↑↓	third transition series
75	rhenium	2	6	10	14	2	6	↑	↑	↑	↑	↑	↑↓	
76	osmium	2	6	10	14	2	6	↑↓	↑	↑	↑	↑	↑↓	
77	iridium	2	6	10	14	2	6	↑↓	↑↓	↑↓	↑↓	↑		
78	platinum	2	6	10	14	2	6	↑↓	↑↓	↑↓	↑↓	↑	↑	
79	gold	2	6	10	14	2	6	↑↓	↑↓	↑↓	↑↓	↑↓	↑	
80	mercury	2	6	10	14	2	6	↑↓	↑↓	↑↓	↑↓	↑↓	↑↓	

Figure 19 *continued.*

There is a notable difference between the metals of the first transition series (first long period) and those of the other two series, which more closely resemble each other. In the first series, the metals are fairly stable in the ionic state as M^{2+} or M^{3+}; they can be oxidized to form compounds of higher valencies which are often oxidizing agents, that is the metal is not so stable at the higher valency. In the other two series, there is little tendency to form simple ionic compounds; the compounds at higher valencies are easily formed and are more stable, that is, they are less likely to be oxidizing agents. This smaller 'gap' between the properties of the second and third long period elements, compared with that between the first and second, is attributed to the lanthanide contraction (see under f-block, below): it leads to the atoms of the third long period elements being smaller than one might expect from their atomic mass; they are almost exactly the same size as the corresponding atoms of the second long period elements (see figure 20). It is not surprising to find atoms which are the same size, and have the same outer electronic structure, showing very similar properties. It will be noted from figure 20 that there is little variation in atomic size amongst the central elements in each period. Many of these metals readily form alloys with each other; some of the metals occur together in ores, and are rather difficult to separate (Zr and Hf; Nb and Ta).

Emission spectra

Transition metals have unusually complicated emission spectra, and it was this fact that gave rise to the name 'd-block' elements, for the d stands for diffuse, applied to the spectra. These spectra are caused by the transitions of electrons from one orbital to another, a process which

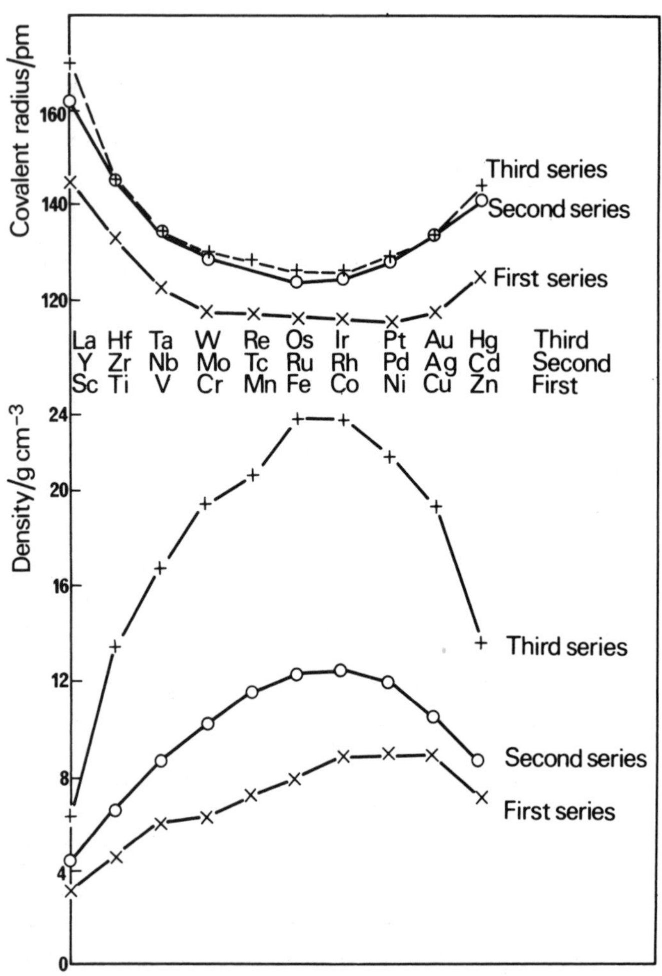

Figure 20 Density and covalent radii of transition metals.

involves the absorption of energy, and then its re-emission. In the case of these metals, there are more available orbitals than for s- or p-block elements, so that there are many more possible transitions.

Absorption spectra

The compounds of these metals, whether ionic or covalent, are coloured in most cases. This is again connected with the number of available orbitals, and the fact that the amount of energy needed to cause an electron to change orbitals is very small: it corresponds to the energy of a photon of light within the visible region so that certain wavelengths of visible light are absorbed, and the compounds appear coloured.

Magnetic properties

For a similar reason, these metals also show special magnetic properties—their ions usually have a high paramagnetic moment. This is a function of the number of unpaired electrons, and figure 19 shows that they usually have several. Paramagnetic ions tend to move in the direction of increasing magnetic flux, a phenomenon which can be demonstrated by comparing accurately the apparent mass of a small sample in the presence of a strong magnet, and the mass of the same sample with the magnet removed.

Catalytic activity

Another special property that the metals often have is catalytic activity. One mechanism of catalysis is connected with the ability of the elements to form complexes in a number of different oxidation states; intermediate compounds can then be formed with one or both reagents.

Most of the properties of transition elements can be found separately in other elements, but non-transition elements do not show all these properties at once, so we are justified in speaking of 'transition properties' as a convenient group term. The zinc group, which completes the block, is not normally regarded as a transition group; the filled d shell cannot be attacked by oxidizing agents, and the s electrons behave rather like the s electrons of the alkaline earth metal group. The metals are always divalent, although they also form a number of complex compounds in oxidation state +2. The ions are not coloured, and the metals are not outstanding catalysts (mercury catalyses the oxidation of naphthalene).

The scandium group, the first group in the d-block, does not show all the transition properties well developed; for instance, there is no variable valency, and less tendency to form complex compounds than later groups show. The spectra show the characteristic patterns, and there is catalytic activity.

The f-block elements

There are two series of f-block elements, known as lanthanides and actinides. In the first, the 4f orbitals are filling, up to the maximum of 14, and in the second, the 5f orbitals are similarly filling. The first elements, which give their names to the series, have no f electrons, but the remaining electronic structure is similar, so that comparison is valid; the f electrons, being screened by two outer groups of electrons, have little effect on the properties of the series as a whole.

The lanthanides

These metals are highly electropositive, and the main ion is M^{3+}; cerium has the alternative valency four, and some others also show alternative valencies. The decrease in ionic (and atomic) radius from lanthanum to lutetium (see table 14) is known as the lanthanide contraction. Note that successive elements at the beginning of a period always show a decrease in size with increasing nuclear charge; here, the main influence is the increase in nuclear charge from +57 to +71 (24.5 per cent) causing the ionic radius to decrease from 106.1 pm to 84.8 pm (20.1 per cent). However, the decrease has taken place through the intervention of the lanthanides, between the second and third periods of d-block (transition) elements. Now yttrium, above lanthanum, has an ionic radius of 90 pm, so that the increase from the second long period to the third is just over 16 pm in this case. This increase has been wiped out by the decrease through the lanthanides, and we find that hafnium, atomic number 72, and the element above it (zirconium, atomic number 40) are almost identical in size. Apart from the lanthanide contraction, we might expect a difference of about 20 pm. This same similarity of size extends to the pairs of elements Nb and Ta, Mo and W, Tc and Re, and so on, right through the long periods, as already noted, and results in very close similarity of properties.

Atomic number	**Element**	**Electron configuration**								**Ionic radius (M^{3+})/pm**
		4s	4p	4d	4f	5s	5p	5d	6s	
57	La	2	6	10		2	6	1	2	106.1
58	Ce	2	6	10	2	2	6	0	2	103.4
59	Pr	2	6	10	3	2	6	0	2	101.3
60	Nd	2	6	10	4	2	6	0	2	99.5
64	Gd	2	6	10	7	2	6	1	2	93.8
71	Lu	2	6	10	14	2	6	1	2	84.8
							Overall decrease			21.3

Table 14 The electron configuration of the f-block elements.

In accordance with Fajans's rule that increase of size assists ionization of a cation, we find that the later lanthanides are less highly basic, less ready to ionize than the earlier ones, and more ready to form covalent complexes. This observation resulted in a much improved method of separation of the lanthanides, which had otherwise proved very intractable. The method is chromatography on a cation-exchange resin, using a complexing agent as eluant to remove the ions, which come off the ion exchange column in inverse order of atomic number, that is, in the order of increasing atomic and ionic radius.

The actinides

The actinides, the second f-block series, start with element number 89 and finish with element number 103, lawrencium. They thus include eleven of the synthetic transuranic elements. These elements show rather more variety of valency than the lanthanides, because of the general closing up of energy levels in successive shells, which permits electron transitions more readily than in the lanthanides. Otherwise, there is a fairly close similarity between corresponding members of the two series. In particular, the chromatographic separation method outlined above works for actinides. This has proved of great value, and has even assisted identification of the far transuranic elements, which are intensely radioactive with very short half-lives. The actinide contraction is clearly similar to the lanthanide contraction, and from element 89 to element 95 the radius of the M^{3+} ion decreases from 111 to 99 pm. (Figures are not yet available for the other elements.)

Exercises

1 Compare the elements from scandium to manganese with regard to the range and stability of oxidation states exhibited, and the nature of the various oxides formed.

2 Show the family relationship of the elements chromium, molybdenum, and tungsten.

3 Nickel, copper, and zinc are the last three elements of the first row of the d-block. To what extent do they show 'transition' properties?

Practical Exercises

1 Collect chlorides of as many elements as possible, in 7.5 × 1 cm sample tubes, and arrange them in the form of the periodic table. (CARE is required in the case of chlorides such as PCl_3 and PCl_5.)

Observe the nature of the chloride: solid, liquid, gas; dry powder, crystalline, deliquescent; fuming solid or liquid (reacts with moisture); colour; in which block of the periodic table is colour found?

2 Study the loss in mass occurring on heating the crystalline chlorides of sodium, calcium, and barium. Quote the results in terms of the mass of water associated with one mole of anhydrous NaCl, $CaCl_2$, or $BaCl_2$.

Sodium chloride is hygroscopic, calcium chloride crystals are deliquescent, barium chloride crystallizes as $BaCl_2{\cdot}2H_2O$.

3 Prepare a saturated solution of copper(II) chloride. Observe the colour. Dilute a sample of the solution and observe the colour. Add concentrated hydrochloric acid to the dilute solution and observe the colour.

This is an example of the equilibrium between covalent and ionic structures; ionization is promoted by dilution (Ostwald's law), and repressed by the common ion effect. Covalent copper is green; Cu^{2+}(aq is blue.

4 Strongly heat a sample of sodium chloride. Repeat with other chlorides such as anhydrous Fe_2Cl_6, $HgCl_2$, CCl_4. (Do not breathe the vapours.)

5 Demonstrate the conductivity or non-conductivity of NaCl solu tion, HCl solution, $HgCl_2$ solution, and CCl_4, by means of a low-voltage supply with electrodes and a torch bulb in the circuit; *OR* measure the conductivities of, say, 0.1 M solutions with a conductivity bridge.

6 Carry out pH titrations of 0.1 M solutions of phosphoric(v) acid sulphuric(vi) acid, and chloric(vii) acid (perchloric acid) with standard alkali. Comment on the relative strengths of these acids.

7 Determine the pH of 0·1 M solutions of sodium phosphate(v) sodium sulphate(vi), and sodium chlorate(vii) (sodium perchlorate) Narrow-range pH papers will be suitable, but a colorimeter or pH mete would be preferable.

Note. The salt of a strong base and a weak acid hydrolyses to give an alkaline solution. Comment on the relative strengths of phosphoric(v) sulphuric(vi), and chloric(vii) acids (see exercise 6).

8 Prepare several cubes of ice by scooping a small hollow with a penknife. Place a small pellet of dried lithium, sodium, and potassium in successive cubes. (Stand back!) Repeat with calcium metal and mag nesium. Take a further sample of calcium and magnesium metal in a test-tube of cold water. Subsequently, heat the test-tube containing magnesium. Comment on your observations.

9 Examine the effects of dilute hydrochloric acid on samples of zinc tin, iron, copper, and aluminium. (Other metals may be tested, if avail able.) Heat if necessary.

10 Examine the effects of dilute sodium hydroxide solution on the metals tested above. Tabulate and discuss the results.

Note. A metal which reacts with both acids and alkalis is described a amphoteric.

11 In the case of metals which react sparingly or not at all with hydrochloric acid, repeat the test with dilute nitric acid.

Note. Nitric acid is an oxidizing agent. A number of transition metals are very susceptible to oxidation. Remember, for instance, the effec of aqua regia on gold, which is otherwise very resistant to attack.

12 *Demonstration*. The properties of chlorine as a very reactive substance; for example attack on Dutch metal, or on metal filings.

Demonstration. The properties of iodine as a very reactive substance

Mix *dry* iodine with powdered aluminium or zinc in small quantities in a *dry* evaporating basin. Add a few drops of water. After a few moments, reaction sets in and heat is generated, which volatilizes some iodine.

13 Show the gradation of halogen properties by adding chlorine water to a solution of a bromide and of an iodide. In each case add a little tetrachloromethane (carbon tetrachloride) and shake. The less active halogen is displaced.

Compare the action of concentrated sulphuric(VI) acid on samples of sodium chloride, bromide, and iodide.

14 Prepare 0.1 M solutions of $MgCl_2$, $CaCl_2$, $SrCl_2$, and $BaCl_2$, and determine their pH (colorimeter or instrumental measurements).

Repeat with 0.1 M solutions of Na_2SO_4, $MgSO_4$, and $Al_2(SO_4)_3$.

Note. The salt of a strong acid and a weak base hydrolyses to give an acidic solution. Hence comment on the relative strengths of the corresponding bases.

Further Reading

Addison, W. E. (1961) *Structural Principles in Inorganic Compounds*, Longman.

Cartmell, E. and Fowles, G. W. A. (1970) *Valency and Molecular Structure*, third edition, Butterworth.

Cooper, D. G. (1968) *The Periodic Table*, fourth edition, Butterworth.

Cotton, F. A. and Wilkinson, G. (1972) *Advanced Inorganic Chemistry*, third edition, Interscience.

Liptrot, G. F. (1971) *Modern Inorganic Chemistry*, Mills and Boon.

Puddephatt, R. J. (1972) *The Periodic Table of Elements*, Oxford University Press.

Rich, R. L. (1965) *Periodic Correlations*, Benjamin.

Sidgwick, N. V. (1960) *The Chemical Elements and Their Compounds*, Volumes I and II, Oxford University Press.

Index